Kalman Filter Recent Advances and Applications

Kalman Filter Recent Advances and Applications

Editor

Beata Akselsen

scitus academics

Kalman Filter Recent Advances and Applications
Edited by **Beata Akselsen**

ISBN: 978-1-68117-267-5
Library of Congress Control Number: 2016937435

www.scitusacademics.com
Box No. 4766, 616 Corporate Way,
Suite 2, Valley Cottage,
NY 10989

Notice

Printed in United States of America on Acid Free Paper

Preface

The Kalman filter is an algorithm estimating a system's measured data status; it was mainly developed by the Hungarian engineer Rudolf Kalman. The filter algorithm is a two-step process: the first step foresees the status of the system and the second step uses noisy measurements to refine the evaluation of the system. There are now several variations to the original Kalman filter. There are many technological applications to the Kalman filter; such as vehicle guidance, navigation and control, especially aircraft and spacecraft, are commonly used. The Kalman filter is also widely used in time series analysis concept in areas such as signal processing and econometrics. Kalman filters are also one of the main topics in the planning and control of robotic motion and are sometimes included in trajectory optimization. These are used to track objects to predict the future location of an object, to account for noise in the location of an object detected and to link multiple objects to their tracks. The output of the Kalman filter is displayed in red circles and in black object detection.

This book provides state of the art developments in the theory and applications of Kalman filters in engineering and science. This book corresponds to developments in Kalman's filtering theory, medical and biological applications, electrical engineering, tracking and positioning systems, and, finally industrial processes and communications.

Table of Contents

Chapter 1

Interactive Kalman Filtering for Differential and Gaussian Frequency Shift Keying Modulation with Application in Bluetooth

Mahdi N. Ali, Mohamed A. Zohdy

Electrical and Computer Engineering Department, Oakland University, Rochester, USA.

ABSTRACT

Some applications are constrained only to implement low cost receivers. In this case, designers are required to use less complex and non-expensive modulation techniques. Differential Quadrature Phase Shift Keying (DQPSK) and Gaussian Frequency Shift Keying (GFSK) can be non-coherently demodulated with simple algorithms. However, these types of demodulation are not robust and suffer from poor performance. This paper proposes a new method to enhance the performance of DQPSK and GFSK using Interactive Kalman Filtering (IKF) technique, in which a one Unscented Kalman Filter (UKF) and two Kalman Filters (KF) are coupled to optimize the demodulated signals. This method consists of simple but very effective algorithms without adding complexity to the demodulators comparing to other very complex methods. UKF is used in this method due to its superiority in approximating and estimating nonlinear systems and its ability to handle non-Gaussian noise environments. The proposed method has been validated by creating a MATLAB/SIMULINK Bluetooth system model, in which the IKF is integrated into the receiver, which implement both DQPSK and GFSK, and run simulation in Gaussian and Non-Gaussian noise environments. Results have shown the effectiveness of this method in optimizing the received signals, and that the UKF outperforms the Extended Kalman Filter (EKF).

Keywords: Interactive Kalman Filtering; Unscented Kalman Filter; Extended Kalman Filter; Differential Quadrature Phase Shift Keying; Gaussian Frequency Shift Keying; Bluetooth

1. INTRODUCTION

Receivers with noncoherent demodulation have simple structures, but they don't have robust performance [1-3]. Simple noncoherent demodulators are attractive in applications where cost is a concern. For example, in Bluetooth protocols [4], noncoherent detection of GFSK and variants of Differential Phase Shift Keying (DPSK) are specified due to their simplicity and low costs. Although these non-optimal techniques are sufficient in low noisy channels, they may suffer performance degradations in very noisy environments.

In order to optimize the performance of the noncoherent demodulation, some complex Maximum Likelihood Sequences Estimator (MLSE) is utilized. An efficient method of implementing this estimation process is through the usage of the very complex Viterbi algorithm that searches the paths through the state trellis for the minimum Euclidean distance path [1]. It was shown that using the Viterbi decoder in Bluetooth receiver achieved very superior performance, but added more complexity, comparing to the noncoherent demodulation that consists of Limited Discriminator with integrating and dump filter [5].

DQPSK is one form of DPSK. It is a memoryless, and a discontinuous phase modulation, in which the current binary sequence is modulated independently of the previous sequences. GFSK is a subclass of Continuous Phase Modulation (CPM), and an attractive modulation in many applications due to its enhanced spectral properties. In modulations, where phase is not continuous, poor spectral efficiency is a real issue. Discontinuities and abrupt changes in phase cause this poor performance, making CPM a more favored scheme [2,6]. GFSK is a modulation with memory, in which the current transmitted signal depends on the current and previous binary sequences. In DQPSK, the phase is constant over a one symbol interval, while in GFSK the phase varies continuously over the symbol period [7]. Both DQPSK and GFSK can be detected non-coherently, which results in simplified transceivers [8].

In the latest Bluetooth specifications, two types of modulation schemes are specified. The first one is the GFSK, which is a form of Continuous Phase Frequency Shift Keying (CPFSK), and hence CPM [9]. It is used in the Basic Data Rate (BDR), with transmission rate at 1 Mbps. The other type is the Differential Phase Shift Keying (DPSK), which is used in the Enhanced Data Rate (EDR) with two variants, $\pi/4$-DQPSK and 8DPSK. The transmission rates are 2 Mbps for EDR using $\pi/4$-DQPSK and 3 Mbps using 8DPSK [4]. Data are transmitted in frames of 625 µs time slots. Each frame is divided mainly into Access Codes (AC) and Payloads. In the EDR, the AC is modulated using one of the DPSK variants, while the payload is modulated using GFSK. No equalization requirements are specified in Bluetooth protocols in order to preserve the low complexity of the system, leaving this area open for researchers to propose efficient equalization and optimization methods that improve the performance of the system, but without inducing much costs.

Significant amount of researches have been conducted to optimize the performance of the DQPSK and GFSK noncoherent demodulation. A. Soltanian and R. E. Van Dyck showed that in typical Bluetooth indoor applications, the noncoherent limiter-discriminator with integrate and dump filter receiver, without equalization, achieves reasonable performance; but this is not the case for the outdoor and large indoor applications, where the signal needs to be more robust. They suggested the usage of Viterbi receiver to get optimum results for such applications [5]. In [9], N. Ibrahim, L. Lampe, and R. Schober presented a receiver design for a GFSK demodulator based on Laurent's decomposition. They developed a Non-coherent Decision Feedback Equalizer (NDFE) and showed that it achieved a robust performance comparing to other equalization methods. In their method, they handled the nonlinearity in the modulation by transforming it into linear using Laurent's decomposition.

M. Nafie, A. Gatherer, and A. Dabak presented a nonadaptive a fractionally sampled Decision Feedback Equalizer (DFE) to be added after the discriminator to enhance the performance of noncoherent GFSK receivers. The proposed DFE was non-adaptive, and they proposed this equalizer to be trained off line at a certain SNR and then this will be used in reception [10]. Harry Leib presented his work that considered an optimal noncoherent demodulation of DPSK when the receiver has partial knowledge of the transmitted data. They proposed A Maximum Likelihood (ML) noncoherent receiver in [11]. The same technique was implemented for GFSK signals in [12].

Implementing Kalman Filtering to enhance the performance of noncoherent continuous and differential phase demodulation have been proposed in several articles. In [13,14], Gal, Campeanu, and Nafornita proposed a method of applying Extended Kalman Filtering in the noncoherent demodulation of CPM and CPFSK respectively. In [15], O. Loffeld utilized the EKF in the demodulation of noisy phase or frequency of sinusoidal signals of unknown amplitude in the presence of an Additive Wideband Gaussian Noise (AWGN). He assumed that the state space model of the unknown phase and amplitude were known. Those Proposed papers utilized linearization technique on the system before using the EKF.

In 1960, Rudolf E. Kalman presented his Kalman Filter method in [16]. KF is an estimator, in which it estimates the states in linear dynamic systems by minimizing the Minimum Mean Squared Error (MMSE). Kalman invented this based on modifying Wiener-Kolmogorov filter, which was derived in 1940 [17]. KF is originated from Baye's probabilistic theory, in which if good number of past samples are known, the future samples can be predicted and updated based on the continuously collected results [18]. Techniques have been proposed to modify KF to be applied to nonlinear systems. For example, EKF has been proposed in nonlinear systems estimations by linearizing the estimated variables through deriving Jacobian matrices. However, EKF may not be a good choice in system with high nonlinearity, or systems that are very difficult to calculate their Jacobian matrices [19].

All the referenced techniques as mentioned thus far are either very complex, or suffer from inadequacy in handling nonlinearly and non-Gaussian noise environments. A better method to approximate nonlinear systems and apply Kalman filtering equations to them was proposed by S. Julier and J. Uhlmann [20]. UKF was derived based on the unscented transformation (UT) concept, which is used to approximate random variables distributions that are going through nonlinear transformations without a need to calculate the probability density functions (pdf) of the nonlinearly transformed variables. UKF has been demonstrated to be very effective in applications when dealing with nonlinear dynamic systems and non-Gaussian types of noises comparing to other filtering techniques [21]. Several types of UKF methods have been developed based on different techniques of UT. These techniques include the General Unscented Transformation, the Scaled Unscented Transformation, the Reduced Sigma Point, and the Spherical Simplex Unscented Transformation [22-25]. In [26], M. Ali and M. Zohdy proposed using UKF in the equalization of CPFSK. They showed that UKF outperformed EKF when both implemented in the CPFSK.

In this paper, we are proposing the implementation of IKF to enhance the performance of the noncoherent demodulation of DQPSK and GFSK. It will be shown that UKF has superior performance in approximating nonlinear systems, and in handling non-Gaussian noise than the already existed EKF methods. This proposed method is an attractive one in the sense that it has simple algorithms so that not that much complexity is added to the receivers comparing to the optimum demodulation. Simulation will be run on a developed MATLAB/SIMULINK model representing Bluetooth system. Results will be compared to those generated by the same model before the implementation of this method, and when using EKF.

This paper has been organized as follow. Introduction is presented in the Section 1. Section 2 will details the new method of the unscented transformation, and how to apply it in UKF. Section 3 will discuss the signal descriptions of DQPSK and GFSK, and Section 4 will detail the system models of both. Section 5 will explain the IKF algorithms. In Section 6, the simulation and results will be listed. Section 7 lists the conclusion.

2. UNSCENTED TRANSFORMATION AND UNSCENTED KALMAN FILTER

The UT does not approximate the nonlinear process and observation functions. It approximates the statistics (mean and covariance) of the state variable x which has dimension n_x by carefully choosing a deterministic set of sample points $\mathcal{X}_i$ and their corresponding weights W_i which have dimensions $2n_x+1$. These sigma points and their associated weights are chosen carefully to capture accurately the actual mean and covariance of the state variables, and when the sampled mean and covariance are propagated through a nonlinear system $z=h(x)$, the a posteriori mean and covariance of the state variables can be estimated accurately up to the 3rd order of any nonlinearity [20]. The statistics of the random variables can also be approximated for higher nonlinearity (4th order and up) but with some errors. **Figure 1** illustrates the principle behind the UT when some deterministic points are undergoing nonlinear transformation.

2.1. Choosing Sigma Points

There is no single method of choosing sigma points that guarantees global optimization of a system. Rather, many methods have been proposed in which each one has advantages and disadvantages over the others. Simon Julier proposed the Scaled Unscented Transformation (SUT) in [23] to be able to control the sigma points without the possibilities of the covariance to become non-positive semi-definite. In the SUT, the sigma points are calculated as follow:

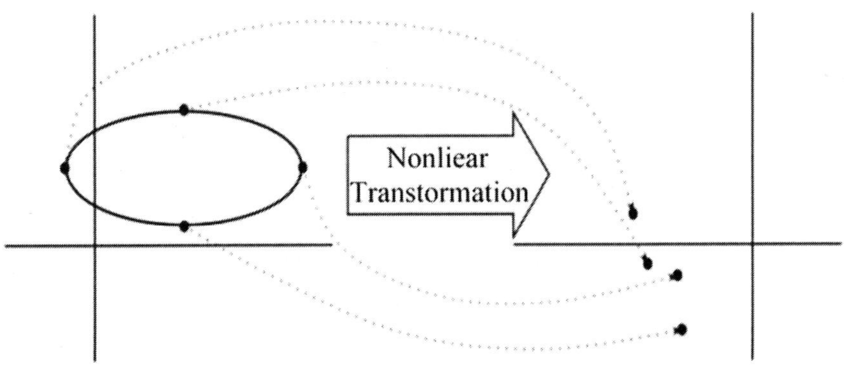

Figure 1. Nonlinear transformations using unscented transformation.

$$
\begin{aligned}
&\text{For } \ i=0: \\
&\mathcal{X}_0=\bar{x}, W_0^m=\frac{\lambda}{n_x+\lambda}, W_0^c=\frac{\lambda}{n_x+\lambda}+\left(1-\alpha^2+\beta\right) \\
&\text{For } \ i=1,\cdots,2n_x: \\
&W_i^m=W_i^c=\frac{1}{2\left(n_x+\lambda\right)} \\
&\mathcal{X}_i=\bar{x}\pm\left(\sqrt{\left(n_x+\lambda\right)P_x}\right)_i
\end{aligned}
\tag{1}
$$

where, $\lambda=\alpha^2\left(n_x+\kappa\right)-n_x$, and W_i^m and W_i^c are the associated weights of these sigma points. The scaling factor λ is a tool to scale the sigma points towards or away from the mean. It should be noted that there is no general way to propose the values of the scaling factors $\beta, \alpha, \text{ and } \kappa$. In each application, these values should be chosen to reduce the estimation errors for their specific problems. As an example, for Gaussian distributions, the scaling factors $\beta, \alpha, \text{ and } \kappa$ can be set to 2, 0.001, and 0 respectively for optimum estimation [21].

2.2. The Unscented Kalman Filtering

The UKF is a novel method that was proposed in [20] to propagate the mean and covariance of a random variable through nonlinear systems using the unscented transformation in order estimate the a posteriori mean and covariance. The following lists the steps for the UKF algorithm based on the SUT.

2.2.1. Nonlinear System Equations

A nonlinear system can be represented as:

$$x_{k+1}=f\left(x_k,u_k,w_k\right) \tag{2}$$

$$z_k=h\left(x_k,v_k\right) \tag{3}$$

Equation (2) models the system process model, where x_k represents the states to be estimated, u_k represents the input, and w_k represents the process noise that has zero mean and covariance Q_k. Equation (3) models the observation model, where v_k represents the measurement noise, which has zero mean and covariance R_k.

2.2.2. Initialization and Sigma Points Calculation

The state vector and process covariance matrix need to be initialized as follow:

$$\hat{x}_0^a = E\left[x_0^a\right] = E\left[x_0 \quad 0 \quad 0\right]^T \tag{4}$$

$$P_0^a = E\left[\left(x_0^a - \hat{x}_0^a\right)\left(x_0^a - \hat{x}_0^a\right)^T\right] \tag{5}$$

where $x_0^a = \left[x_0 \quad w_0 \quad v_0\right]^T$. Equation (1) can be used to derive $2n_x + 1$ deterministic sigma points, which can be written as:

$$\mathcal{X}_{k-1}^a = \left[\hat{x}_{k-1}^a \;\; \hat{x}_{k-1}^a + \sqrt{\left(n_x + \lambda\right) P_{k-1}^a} \;\; \hat{x}_{k-1}^a - \sqrt{\left(n_x + \lambda\right) P_{k-1}^a}\right] \tag{6}$$

2.2.3. Time Update Equations

The sigma points in Equation (6) are propagated through the nonlinear process function in Equation (2) to give:

$$\mathcal{X}_{k|k-1}^a = f\left(\mathcal{X}_{k-1}^a, u_k\right) \tag{7}$$

The sampled predicting mean and covariance of the state variable are calculated in the next two equations.

$$\hat{x}_{k|k-1} = \sum_{i=0}^{2n_x} W_i^m \mathcal{X}_{i,k|k-1}^a \quad (8)$$

$$P_{k|k-1} = \sum_{i=0}^{2n_x} W_i^c \left[\left(\mathcal{X}_{i,k|k-1}^a - \hat{x}_{k|k-1} \right) \left(\mathcal{X}_{i,k|k-1}^a - \hat{x}_{k|k-1} \right)^T \right]. \quad (9)$$

2.2.4. Measurement Update Equations

In order to derive the measurement update equations, the sigma-point states as derived in Equation (7) need to be propagated in the nonlinear measurement model in Equation (3).

$$\mathcal{Y}_{k|k-1} = h\left(\mathcal{X}_{k|k-1}^a, v_{k-1} \right) \quad (10)$$

The output sampled mean and covariance are captured as follow:

$$\hat{y}_{k|k-1} = \sum_{i=0}^{2n_x} W_i^m \mathcal{Y}_{i,k|k-1} \quad (11)$$

$$P_y = \sum_{i=0}^{2n_x} W_i^c \left[\left(\mathcal{Y}_{i,k|k-1} - \hat{y}_{k|k-1} \right) \left(\mathcal{Y}_{i,k|k-1} - \hat{y}_{k|k-1} \right)^T \right] \quad (12)$$

The cross covariance of the state and measurement can be calculated as:

$$P_{x,y} = \sum_{i=0}^{2n_x} W_i^c \left[\left(\mathcal{X}_{k|k-1}^a - \hat{x}_{k|k-1} \right) \left(\mathcal{Y}_{i,k|k-1} - \hat{y}_{k|k-1} \right)^T \right] \quad (13)$$

Using the conventional KF equations, we can calculate Kalman gain as:

$$K_k = P_{x,y} \cdot P_y^{-1} \tag{14}$$

where, P_y and $P_{x,y}$ are derived in Equations (12) and (13) respectively. The Kalman gain as derived in Equation (14) can be used in the following equation to calculate the updated a posteriori mean of the states.

$$\hat{x}_k = \hat{x}_{k|k-1} + K_k \left(y_k - \hat{y}_{k|k-1} \right) \tag{15}$$

Finally, the update covariance can be expressed as:

$$P_k = P_{k/k-1} - K_k P_y K_k^T \tag{16}$$

3. Π/4 DQPSK AND GFSK SIGNAL DESCRIPTION

3.1. π/4 DQPSK Signal Description

The π/4-DQPSK is a version of DQPSK, but shifted by π/4. For simplicity we refer to π/4-DQPSK as DQPSK throughout this paper. Its transmitted signal can be represented as [27]:

$$s_i(t) = \sqrt{2E_s/T_s} \cos\left(2\pi f_c t + \theta_i(t)\right), \; 0 \le t \le T_s \tag{17}$$

where $s_i(t)$ represents the carrier-modulated signal, $\theta_i(t)$ is the information-carrying phase that mapped to a symbol i , E_s and T_s are the symbol energy and duration respectively. $i = 1, 2, \cdots, M$ and M is the number of symbols, which is 4 in the case of DQPSK. At $t = k$ the dynamic behavior of the phase can modeled as:

$$\theta_k = \left(\theta_{k-1} + \Delta\phi_k\right)_{\text{mod } 2\pi} \tag{18}$$

where the addition is modulo 2π. $\Delta\phi_k$ is the value, which results from mapping the binary sequences to the phase representation. Thus, 11, 01, 00, 10 sequences are mapped to $\pi/4, 3\pi/4, -3\pi/4, -\pi/4$ respectively. Equation (18) shows the dynamic transition when phase is changing from current transmitted symbol to the next one. The Inphase I_k and Quadrature Q_k components (complex envelops) of the transmitted signal of the kth symbol are determined as follow:

$$I_k = \cos(\theta_k) = I_{k-1}\cos(\Delta\phi_k) - Q_{k-1}\sin(\Delta\phi_k) \tag{19}$$

$$Q_k = \sin(\theta_k) = I_{k-1}\sin(\Delta\phi_k) + Q_{k-1}\cos(\Delta\phi_k) \tag{20}$$

The Inphase I_k and Quadrature Q_k are functions of the phase θ_k which is in turn functions of the previous state θ_{k-1} and the current input $\Delta\phi_k$. This is modulated this way so that the only information needed at the demodulator side is the difference between the consecutive received phases, which enable simpler noncoherent demodulation. The In-phase and Quadrature are passed through pulse shape filters to smooth the pulses and conserve the bandwidth. When pass through these filters, the resulted pulses are represented as:

$$I(t) = \sum_{k=0}^{N-1} I_k P\left(t - kT_s - \frac{T_s}{2}\right) = \sum_{k=0}^{N-1} \cos(\theta_k) P\left(t - kT_s - \frac{T_s}{2}\right) \tag{21}$$

$$\begin{aligned} Q(t) &= \sum_{k=0}^{N-1} Q_k P\left(t - kT_s - \frac{T_s}{2}\right) \\ &= \sum_{k=0}^{N-1} \sin(\theta_k) P\left(t - kT_s - \frac{T_s}{2}\right) \end{aligned} \tag{22}$$

There are several forms of the pulse shapes $P(t)$, which can take the forms of Gaussian, Raised Cosine, and others. When these components are multiplied by the carrier, the resulted baseband modulated signal is:

$$S(t) = I(t)\cos(2\pi f_c t) - Q(t)\sin(2\pi f_c t) \tag{23}$$

3.2. GFSK Signal Description

The transmitted carrier-modulated GFSK signal can be represented in the following format [12,27]:

$$s(t,a) = \sqrt{2E_s/T_s}\cos\left(2\pi f_c t + \theta(t,a)\right) \tag{24}$$

The complex baseband waveform of this signal can be expressed as:

$$v(t) = \sqrt{2E_s/T_s}\, e^{i\theta(t,a)} \tag{25}$$

where the phase can be represented as:

$$\theta(t,a) = \pi h \sum_{k=-\infty}^{n-L} a(k) + 2\pi h \sum_{k=n-L+1}^{n} a(k)q(t-kT_s) \tag{26}$$

where $nT_s \le t \le (n+1)T_s$, $s(t,a)$ represents the carriermodulated signal, $\theta(t,a)$ is the time-varying phase of the carrier, E_s is the symbol energy, T_s is the symbol duration, and f_c is the carrier frequency. The digital sequence a_k represents the M-ary information symbols that can take the following symbol format $\pm 1, \pm 2, \pm 3, \dots, \pm(M-1)$, and for binary sequences, $a_k = \{-1,1\}$. h is the modulation index. The first term on the right side of Equation (26) models the phase history. $q(t)$ is the phase response, which is derived by integrating a pulse $g(t)$, and L is the length of this pulse (per unit T_s). $g(t)$ is defined as the instantaneous frequency of $q(t)$. The shape of $g(t)$ determines the smoothness of the transmitted carrying information phase [28]. For example, in the case of GFSK, $g(t)$ is shaped using Gaussian pulses instead of rectangular as this the case in CPFSK. There are Different pulse shapes, such as Raised Cosine (RC), and Gaussian Minimum Shift Keying (GMSK), etc.

4. Π/4 DQPSK AND GFSK SYSTEM DESCRIPTION

In Kalman Filtering, two steps are performed; prediction and update. It is important that the received signal to be predictable in order to use Kalman Filters. Thus, the system under study needs to be modeled to allow performing prediction. Both DQPSK and GFSK dynamic systems will be modeled nonlinearly, and transformed into state space format so that Kalman filtering can be applied. In the noncoherent DQPSK, the current phase is a function of the previous phase and the current input, making it possible to do the prediction task of the next phase. In the case of GFSK, or any other CPM, Kalman Filtering can be an attractive technique since the transition in phase is continuous, and the changes do not occur abruptly, making the prediction step possible also.

4.1. π/4 DQPSK Dynamic System Description

Before the signal goes through decoding and decision state, it goes through the proposed Interactive Kalman Filtering for optimization. In order to apply Kalman Filtering, we need first to derive the state space equations that describe the system. The RC shaping pulse $g(t)$ causes the phase response $q(t)$ to be nonlinear. Therefore, at $t=k$, the phase can be represented nonlinearly using Euler's formula as:

$$\theta(t,a)=\theta_k=\theta_{k-1}+h\cdot f\left(\theta_{k-1}\right) \tag{27}$$

where the delta change $h\cdot f\left(\theta_{k-1}\right)$ representing the change from one phase to the next one can be approximated as:

$$\left(\Delta\phi_{k-1}\right)^2=\theta_k-\theta_{k-1}+w_k \tag{28}$$

Equation (28) is the first order difference equation of the phase. $(\Delta\phi_{k-1})^2$ is a discrete value and constant over one symbol duration. a represents the symbol sequences, and w_k is the process noise, which is assumed to have zero mean and variance σ_w^2. h is the step size, which is treated as a fixed number in this paper, that determines the change value from $k-1$ to k. The smaller this value is, the smaller the nonlinear approximation error is. $f(\theta_{k-1})$ is the rate of the change (slope) at $t=k-1$. If a rectangular pulse shape $g(t)$ was considered as it is the case in CPFSK, than equation (27) can be modeled linearly as $\theta(t,a)|_{t=k}=\theta_k=\theta_{k-1}+\Delta\phi_{k-1}$ [13]. However, for GFSK, this behavior is nonlinear, and Equation (27) can be represented as:

$$\theta(t,a)|_{t=k}=\theta_k=\theta_{k-1}+(\Delta\phi_{k-1})^2 \quad (29)$$

4.1.1. State Transition Model

There are two states in the system, the phase θ_k and the difference equation $\Delta\phi_k$ that need to estimated. The following equations represent the dynamic system of the DQPSK in state space format.

$$\theta_k=\theta_{k-1}+(\Delta\phi_{k-1})^2 \quad (30)$$

$$\Delta\phi_k=\Delta\phi_{k-1}+w_k \quad (31)$$

These can be written in the following matrix format:

$$\begin{bmatrix} \theta_k \\ \Delta\phi_k \end{bmatrix} = \begin{bmatrix} \theta_{k-1} + (\Delta\phi_{k-1})^2 \\ \Delta\phi_{k-1} \end{bmatrix} + \begin{bmatrix} 0 \\ 1 \end{bmatrix} w_k \tag{32}$$

In order to calculate $(\Delta\phi_{k-1})^2$, some a priori knowledge of θ_k at $t = k$ is needed. Thus, Equation (28) can be re-written as:

$$(\Delta\phi_{k-1})^2 = \theta_{k|k-1} - \theta_{k-1} + w_k \tag{33}$$

where $\theta_{k|k-1}$ is the prediction of the phase at $t = k$ having knowledge of the phase up to $t = k-1$. $\theta_{k|k-1}$ can be estimated using two coupled Kalman Filters, as it will be explained in Section 5.

4.1.2. Measurement Models

When transmitted through an ideal channel, which is corrupted by AWGN, the Inphase and Quadrature components of the received signal at $t = k$ can be represented as:

$$y_k = h(\theta_k) + v_k = \sqrt{2E_s/T_s} \cdot \begin{bmatrix} \cos(\theta_k) \\ \sin(\theta_k) \end{bmatrix} + \begin{bmatrix} v_{I_k} \\ v_{Q_k} \end{bmatrix} \tag{34}$$

where v_{I_k} and v_{Q_k} are the measurement noise of the Inphase and Quadrature components respectively.

4.2. GFSK Dynamic System Description

GFSK is different from the DQPSK in a sense that it is a frequency shift keying modulation, and its phase is changing continuously during a symbol transmission [7]. There are two states to estimate in this system, the continuously changing phase $\theta(t, a_n)$ and the difference equation $\Delta\phi(t, a_n)$, where a_n is the input sequences. The dynamic equations can be represented as:

$$\theta(t,a_n)_k = \theta(t,a_n)_{k-1} + h \cdot f(\theta(t,a_n)_{k-1} \\ kT \le t \le (k+1)T \tag{35}$$

$$\left(\Delta\phi(t,a_n)_{k-1}\right)^2 = \theta(t,a_n)_k - \theta(t,a_n)_{k-1} + w_k \\ kT \le t \le (k+1)T \tag{36}$$

Equation (35) states that when the state changes from $k-1$ to k over a symbol a_n with duration $kT \le t \le (k+1)T$, $\theta(t,a_n)_k$ can be determined using Euler's formula by adding the change in the phase $h \cdot f\left(\theta(t,a_n)_{k-1}\right)$ to the previous phase $\theta(t,a_n)_{k-1}$. h is the step size, and w_k is the process noise. This dynamic system can be modeled nonlinearly as:

$$\theta(t,\boldsymbol{a}_n)_k = \theta(t,\boldsymbol{a}_n)_{k-1} + \left(\Delta\phi(t,\boldsymbol{a}_n)_{k-1}\right)^2 \tag{37}$$

At the receiver, the signal can be modeled as:

$$r(t) = \sqrt{\frac{2E_s}{T_s}} \mathrm{e}^{i\left(\theta(t,a_n)_{k-1} + \left(\Delta\phi(t,a_n)_{k-1}\right)^2\right)} + v(t) \tag{38}$$

where $kT_s \le t \le (k+m-1)T_s$.

4.2.1. State Transition Model

The dynamic system of the phase and difference equation of the GFSK can be written as:

$$\theta(t,a_n)_k = \theta(t,a_n)_{k-1} + \left(\Delta\phi(t,a_n)_{k-1}\right)^2 \tag{39}$$

$$\Delta\phi(t,a_n)_k = \Delta\phi(t,a_n)_{k-1} + w_k \tag{40}$$

In a matrix format, (39) and (40) can be represented as

$$\begin{bmatrix} \theta(t,a_n)_k \\ \Delta\phi(t,a_n)_k \end{bmatrix} = \begin{bmatrix} \theta(t,a_n)_{k-1} + \left(\Delta\phi(t,a_n)_{k-1}\right)^2 \\ \Delta\phi(t,a_n)_{k-1} \end{bmatrix} + \begin{bmatrix} 0 \\ 1 \end{bmatrix} w_k \tag{41}$$

4.2.2. Measurement Model

When transmitted through an ideal channel with AWGN, the Inphase and Quadrature components of the received signal at $t = k$ can be represented as:

$$y_k = h(\theta_k) + v_k = \sqrt{2E_s/T_s} \cdot \begin{bmatrix} \cos(\theta_k(t,a)) \\ \sin(\theta_k(t,a)) \end{bmatrix} + \begin{bmatrix} v_{I_k} \\ v_{Q_k} \end{bmatrix} \tag{42}$$

4.3. π/4 DQPSK and GFSK Dynamic System Description

There are communication systems that implement multiple modulation schemes, such as WIFI, Bluetooth, etc. **Figure 2** represents a receiver that implements both DQPSK and GFSK, similar to the modulation schemes as specified in Bluetooth communications [4]. This figure depicts a functional blocks of the demodulator along with the proposed IKF interface. The DQPSK part of the demodulator processes the AC and Headers portion of the frames, while the GFSK part of the demodulator processes the Payloads portion of the frames. As it will be explained in the next section, the two KFs calculate the a priori phase $\theta_{k+1|k}$ and feed it into the UKF. In its turn, UKF uses this a priori and knowledge of the noise characteristics, and apply its simple algorithms to estimate the received signal components, before it feeds it into the decoding and decision devices for further processing.

Next, the state space equations for both DQPSK and GFSK systems are presented, which include 4 states, and 4 outputs.

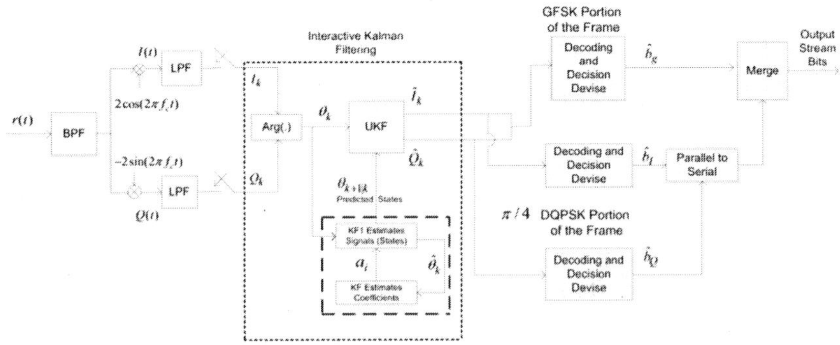

Figure 2. The baseband DQPSK and GFSK receiver with the integrated Kalman filtering.

4.3.1. State Transition Model

$$\begin{bmatrix} \theta_k \\ \Delta\theta_k \\ \theta_k(t,a) \\ \Delta\theta_k(t,a) \end{bmatrix} = \begin{bmatrix} \theta_{k-1} + (\Delta\phi_{k-1})^2 \\ \Delta\phi_{k-1} \\ \theta(t,a_n)_{k-1} + (\Delta\phi(t,a_n)_{k-1})^2 \\ \Delta\phi(t,a_n)_{k-1} \end{bmatrix} + \begin{bmatrix} 0 \\ w_{DQPSK} \\ 0 \\ w_{GFSK} \end{bmatrix}_k \tag{43}$$

4.3.2. Measurement Model

The measurement equations can be written as follow:

$$y_k = h(x_k) + v_k$$
$$= \sqrt{\frac{2E_s}{T_s}} \cdot \begin{bmatrix} \cos(\theta_k) \\ \sin(\theta_k) \\ \cos(\theta_k(t,a)) \\ \sin(\theta_k(t,a)) \end{bmatrix}_k + \begin{bmatrix} v_{I_{DPSk}} \\ v_{Q_{DPSk}} \\ v_{I_{GFSK}} \\ v_{Q_{GFSK}} \end{bmatrix}_k \tag{44}$$

5. INTERACTIVE KALMAN FILTERING

Dual estimations concept has been explored in several researches in different applications. In [29], D. Labarre et al. proposed to take advantage of the so-called instrumental variable (IV) techniques to estimate both the received signal and the associated parameters. They proposed using two conditionally linked Kalman Filters in order to avoid using an EKF to estimate both states and parameters. The first KF uses a new observation to estimate the incoming signal, while the second KF uses this estimated signal to estimate the coefficient parameters. In [30], A. Jamoos, A. Abdo, and H. Nour used similar technique to jointly estimate channel coefficients and parameters using two coupled Kalman Filters in the estimation of rapidly time-varying Rayleigh fading channels in Orthogonal Frequency Division Multiplexing (OFDM) mobile wireless system. Thus, each time a new observation is available, the first filter uses the latest estimated parameters to estimate the signal, while the second filter uses the estimated signal to update the parameters.

5.1. Phase Prediction

This joint estimation method will be utilized to predict the a priori phase $\theta_{k|k-1}$ based on the past measured received signals $\theta_{k-1}, \theta_{k-2}, \cdots$, and feed this predicted value into in UKF to use it in its algorithm. From Equation (29), the phase θ_k is predicted based on the previous value θ_{k-1} and the change as triggered by the current input, which is equivalent to $(\Delta\phi_{k-1})^2 = \theta_k - \theta_{k-1}$. Thus, in order to calculate $(\Delta\phi_{k-1})^2$, we need to have knowledge of θ_k. At the mean time, we need to know $(\Delta\phi_{k-1})^2$ in order to calculate θ_k. The dual estimation is proposed to handle this case to predict the phase θ_k. The phase θ_k is replaced by $\theta_{k|k-1}$ in Equation (29), which can be written as:

$$(\Delta\phi_k)^2 = \theta_{k|k-1} - \theta_{k-1} \tag{45}$$

This a priori can be predicted recursively using the following formula:

$$\theta_{k/k-1} = -\sum_{i=1}^{p} a_i\theta_{k-i} + w_k \tag{46}$$

where a_i is the prediction coefficients, w_k is the so called driving process and it is assumed to be zero-mean noise with variance σ_w^2, and p is the order. This is called joint estimation [18], since both the phase θ_k and their coefficients a_i need to be estimated, and they depend on each others. Two KFs are proposed in order to perform the estimation, which allow the analysis to be run linearly and avoid using any of the nonlinear approximation methods.

KF1 estimates the phases, while KF2 estimates their coefficients. **Figure 3** shows a layout of this proposed IKF method.

5.2. Estimation of the Phase Values

Let $\theta_k = \left[\theta_k \quad \theta_{k-1} \quad \cdots \quad \theta_{k-p-1}\right]^T$. Equation (46) can be put in the following state space format:

$$\theta_k = \Phi_k\theta_{k-1} + gw_k \tag{47}$$

$$y_k = H\theta_k + v_k \tag{48}$$

When $p = 4$, these matrices are defined as:

$$\Phi_k = \begin{bmatrix} -a_1 & -a_2 & -a_3 & -a_4 \\ 1 & 0 & 0 & 0 \\ 0 & 1 & 0 & 0 \\ 0 & 0 & 1 & 0 \end{bmatrix}, \quad g = \begin{bmatrix} 1 \\ 0 \\ 0 \\ 0 \end{bmatrix},$$

$$and \ H = \begin{bmatrix} 1 & 0 & 0 & 0 \end{bmatrix}$$

The goal is to estimate the phase $\hat{\theta}_{k|l}$ at $t = k$, given l observations of $y_1, y_2, \cdots, y_l$ as well as estimating the output $H\hat{\theta}_k$ also. The a posteriori $\hat{\theta}_{k|k}$ is defined as:

$$\hat{\theta}_{k|k} = \Phi_k \theta_{k-1|k-1} + K_k r_k \tag{49}$$

where, r_k is called innovation process and is defined as $r_k = y_k - H\Phi_k \theta_{k-1|k-1}$. Its covariance can be defined as:

$$C_k = H^T P_{k|k-1} H + \sigma_w^2 \tag{50}$$

The so called a priori error covariance matrix $\boldsymbol{P}_{k|k-1}$ can be calculated recursively as:

$$P_{k|k-1} = \Phi_k \theta_{k-1|k-1} \Phi_k^T + g\sigma_w^2 g \tag{51}$$

$\boldsymbol{K}_k$ is called the Kalman gain, and is calculated as:

$$K_k = P_{k|k-1} H^T C_k^{-1} \tag{52}$$

The so called a posteriori covariance is updated as follow:

$$P_{k|k} = \left(I_k - K_k H\right) P_{k|k-1} \tag{53}$$

Finally, the output of KF1 is the predicted phase and can be expressed as:

$$\hat{\theta}_k = H\hat{\theta}_{k|k} \tag{54}$$

The estimated phase $\hat{\theta}_{k|k}$ will be fed into KF2 as the observed values and used in the coefficients estimation, and $\hat{\theta}_k$ will be used as the a priori $\hat{\theta}_{k/k-1}$ and fed into the UKF. The state vector and its covariance can be initialized as $\hat{\theta}_0 = 0$ and $P_0 = I$.

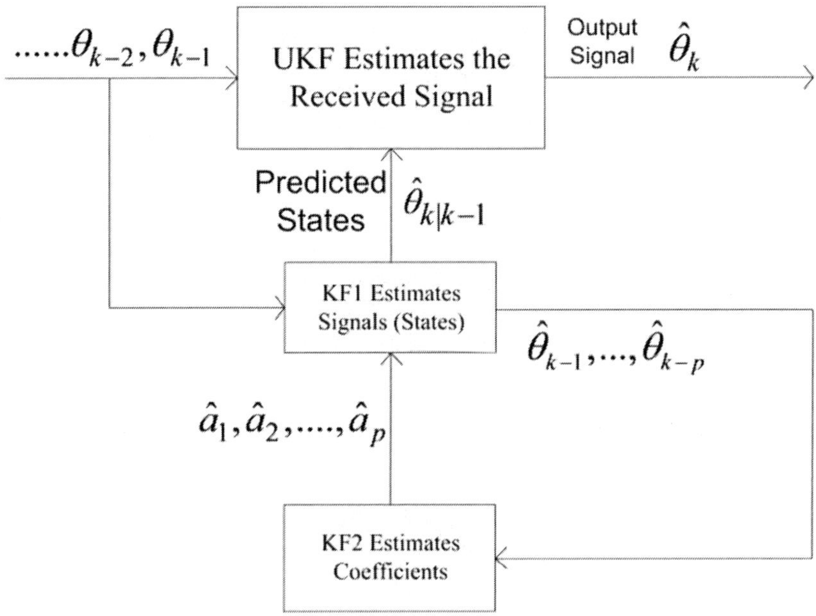

Figure 3. Interactive Kalman filtering.

5.3. Estimation of the Coefficients

The state vector $\hat{\theta}_k$ that was estimated in KF1 is used as the observed value to KF2. In order to estimate the coefficients from the estimated phase, Equations (49) and (54) are combined as:

$$\hat{\theta}_k = H\Phi_k\theta_{k-1} + HK_k r_k = \theta_{k-1}^T a_n + \upsilon_k \tag{55}$$

For the 4th order system, the phase and coefficients vectors are defined as $\hat{\theta}_{k-1} = [\theta_{k-1} \ \ \theta_{k-2} \ \ \theta_{k-3} \ \ \theta_{k-4}]^T$ and $a_n = [-a_1 \ \ -a_2 \ \ -a_3 \ \ -a_4]^T$ respectively. Assuming that the phase signal is stationary or changing very slowly from the current value to the next one, it is possible to assume that the coefficients to be approximately time invariant over a short period of time. In this case they can be written as:

$$a_n = a_{n-1} \tag{56}$$

Now we are ready to define the state space equations for KF2 to estimates the coefficients as:

$$a_n = a_{n-1} \tag{57}$$

$$\hat{\theta}_k = \hat{\theta}_{k-1}^T a_n + \upsilon_k \tag{58}$$

where θ_{k-1}^T become the observed values, and a_n are the states to be estimated.

The covariance of the process υ_k can be calculated as:

$$\sigma_{\upsilon_k}^2 = HK_kC_kK_k^TH^T \tag{59}$$

The coefficients can be recursively computed as:

$$\hat{a}_{k/k} = \hat{a}_{k-1/k-1} + K_k^a\left(\hat{\theta}_k - \hat{\theta}_{k-1/k-1}^T \hat{a}_{k-1/k-1}\right) \tag{60}$$

where the Kalman gain K_k^a and the update state covariance matrix P_k^a can be calculated as:

$$K_k^a = P_{k-1/k-1}^a \hat{\theta}_{k-1}^T \left(\hat{\theta}_{k-1} P_{k-1/k-1}^a \hat{\theta}_{k-1}^T + \sigma_k^2\right)^{-1} \tag{61}$$

$$P_{k/k}^a = \left(I_k - K_k^a \hat{\theta}_{k-1}\right) P_{k/k-1}^a \tag{62}$$

In the same manner, the initial state and its covariance can be initialized as: $\hat{\theta}_0 = 0$ and $P_0 = I$.

6. SIMULATION AND RESULTS

6.1. MATLAB/SIMULINK Bluetooth Model

A Bluetooth model has been created to implement and validate the proposed method. This model was created in the MATLAB/SIMULINK environments [4]. The transmitted signal power is set to 100 mw representing Bluetooth class 1 devices, which can communicate up to 100 meters. The model consists of three parts; the transmitter, the channel, and the receiver. In the transmitter, the data go through the source and channel coding. The data are structured in frames. In the simulation runs, data representing actual speech signals have been processed. Data are organized in frames. Each frame contains 72 bits for Access Codes (AC), 56 bits for Headers, and 240 bits for the payloads, which makes up a total of 366 bits per frame. The 126 bits representing the AC and Headers are modulated using $\pi/4$ DQPSK in this model, and the 240 bits are modulated with GFSK, and transmitted over an AWGN channel in 625 μs time slots. In some of the simulation runs, non-Gaussian channel was used.

The receiver was simulated so that the frame was divided into two sections. The first section is processed by the $\pi/4$ DQPSK portion of the demodulator (bits 1 - 126), while the other sections was build with the GFSK demodulator to process bits 127 - 366. The rest of the blocks in the receiver were simulated to perform all required decoding, and reconstructing of the source signal. The proposed IKF was integrated into the receiver to estimate and optimize the received signals.**Figure 4** shows a high level schematic of this model.

At the receiver, the corrupted signal goes through the IKF with UKF or EKF to perform the estimation steps on it before feeding it into the demodulator. **Figure 5** shows schematic of the integrated IKF, which include the 2 KFs and a UKF as implemented in the receiver. It also includes the blocks to calculate the un-filtered phase and the UKF estimated phase, and their associated MSE.

6.2. Results

Simulation was run to compare the performance of the system with the integrated IKF versus the same system model without Kalman Filtering. Also, the performances of both the UKF and EKF were compared. **Figure 6** shows the results when 200 samples of the received signal were plotted for non-filtered versus the IKF estimated signals, which consists of two KFs and one UKF. The non-noisy transmitted signal was used as a reference. Plots show clearly that IKF improves the signals significantly over the non-enhanced receiver. In the simulation, the SNR was set at value 10. This result shows the advantages of using simple algorithm like IKF to improve the results.

Figure 7 shows the corresponding Square Error (SE), of the un-processed (non-filtered) received signal versus the SE of the IKF estimated signal. These SE values were calculated in reference to the transmitted (non-noisy) signals for the same 200 samples as in **Figure 6**. Results have been consistent in showing that the implementation of IKF has improved the received signal significantly and consistently in enhancing the noncoherent modulation performance.

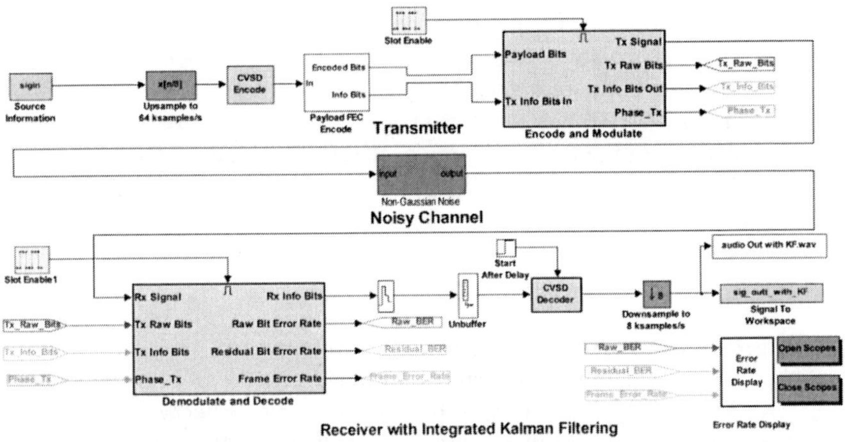

Figure 4. Bluetooth system model with the integrated interactive Kalman filtering in the demodulator.

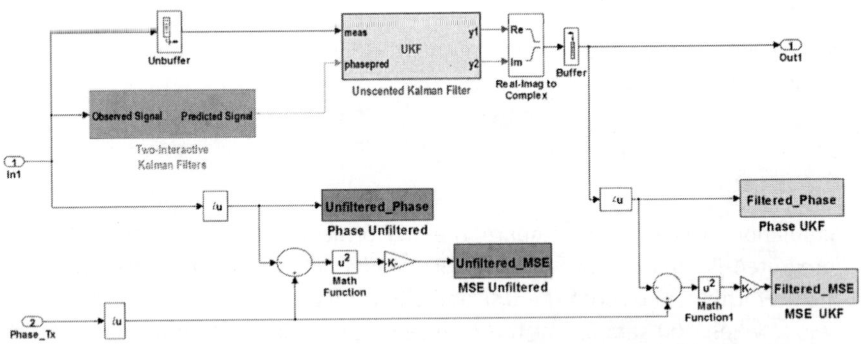

Figure 5. The integrated Kalman filtering as implemented in the bluetooth receiver model.

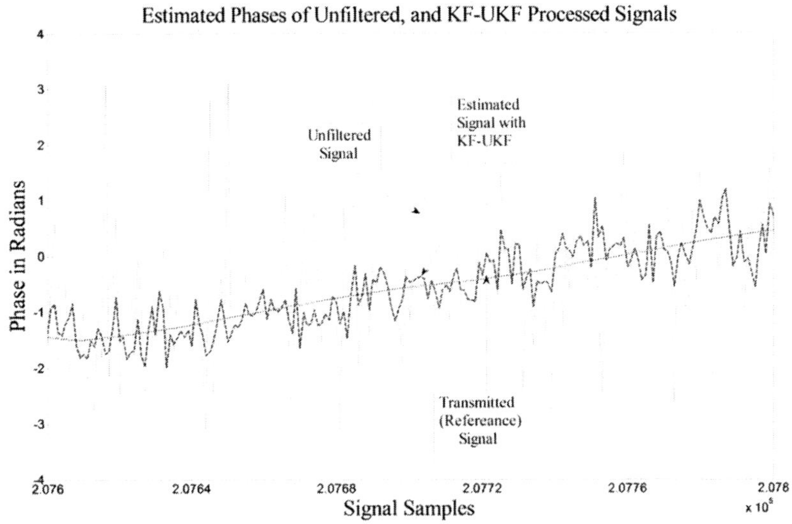

Figure 6. Non-filtered signal versus IKF estimated signals.

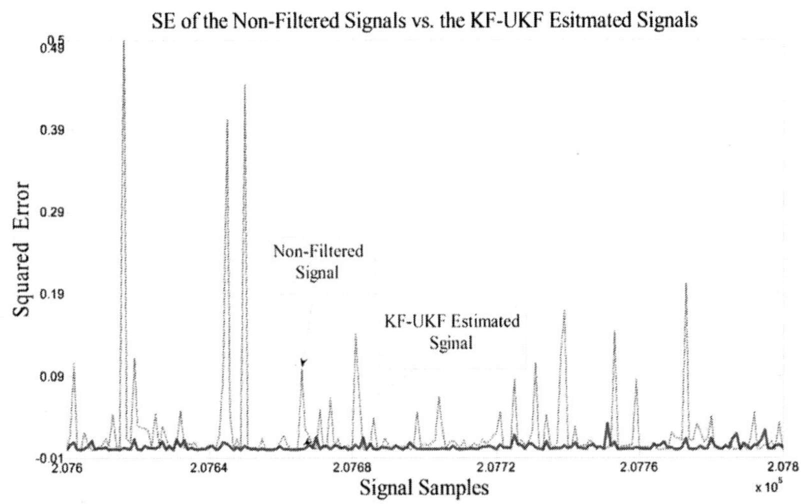

Figure 7. Square error of the non-filtered and IKF estimated signals.

Figure 8 shows the normalized MSEs of the non-estimated, estimated EKF, and UKF signals, when both filters implemented using the proposed IKF method in Bluetooth systems with DQPSK & GFSK modulation in the presence of Gaussian noise. The results show that the both EKF and UKF improve the signals significantly when comparing the results to the receivers with no filtering. It also shows that UKF outperforms the EKF for all SNR values from 0 to 12.

Non-Gaussian noise distributions can be modeled as a mixture of additive zero-mean Gaussians distributions [31]. In [32], L. Binh used SIMULINK to simulate nonGaussian noise as a mix of Gaussian distributions. **Figure 9** shows that UKF continues to outperform the EKF in the presence of non-Gaussian noise, which was created by adding three zero mean Gaussian noise distributions using SIMULINK built-in blocks.

When the step size h in Equations (29) and (35) became smaller, modeling errors would be smaller, enabling modeling the system more accurately for Kalman Filters, which results in better estimation. This is an expected behavior, since with smaller step size the change from the current sample to the next one will be smaller, enabling KF to do more accurate prediction and update. Controlling this processing step size at the demodulator can be done by taking more or less samples per symbol. When taking more samples over the duration of the symbol, this results in smaller step size. **Figure 10** shows clearly the comparison of the BER when the UKF was used in the IKF with 0.01 versus when this step size was doubled to be 0.02. Results show that choosing smaller step size results in much better performance. However, the disadvantage of smaller step size is that more computations are needed.

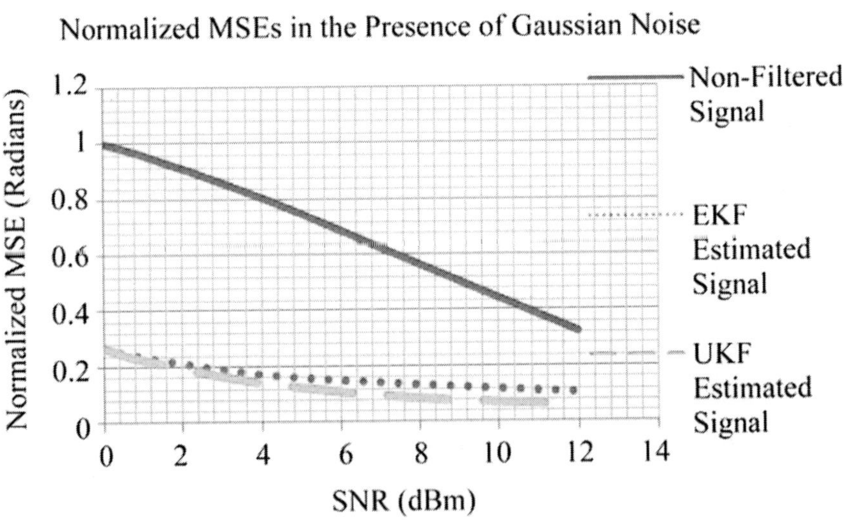

Figure 8. Normalized MSEs of the non-estimated, estimated EKF, and estimated UKF in the presence of Gaussian noise.

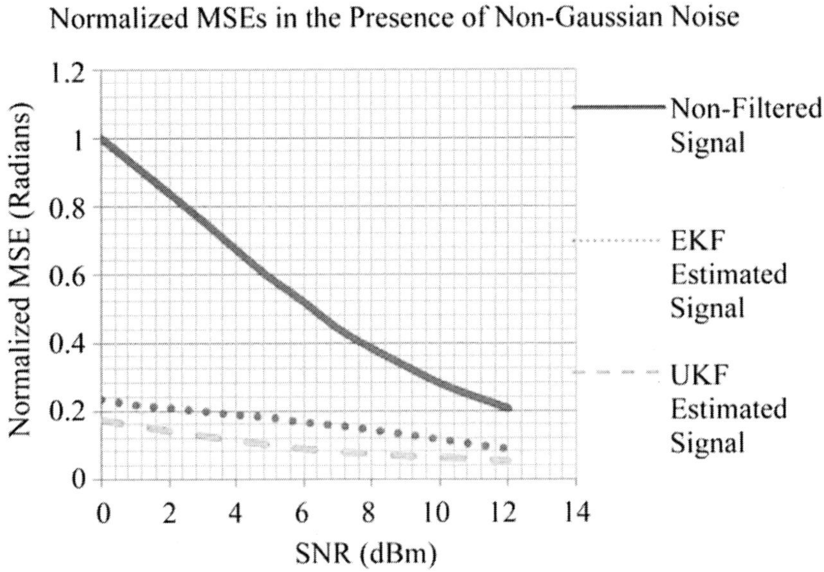

Figure 9. Normalized MSEs of the non-estimated, and estimated EKF and UKF in the non-Gaussian noise.

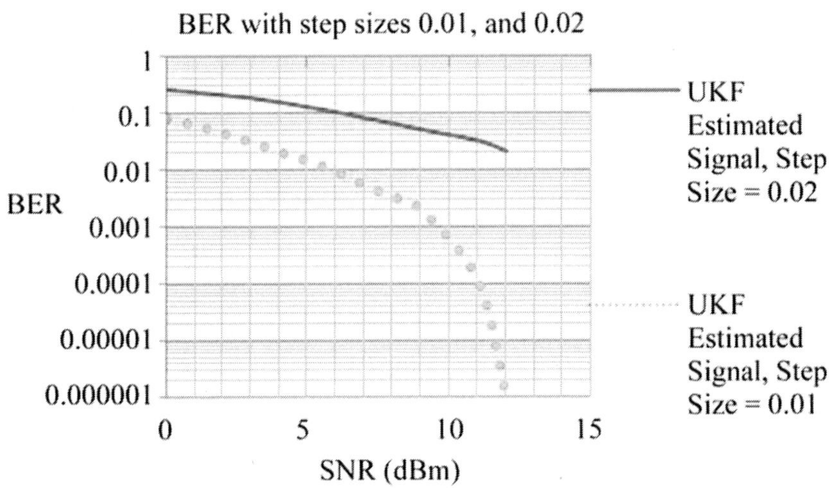

Figure 10. BER of the DQPSK and GFSK demodulated signal when step size was chosen to be 0.01 and 0.02.

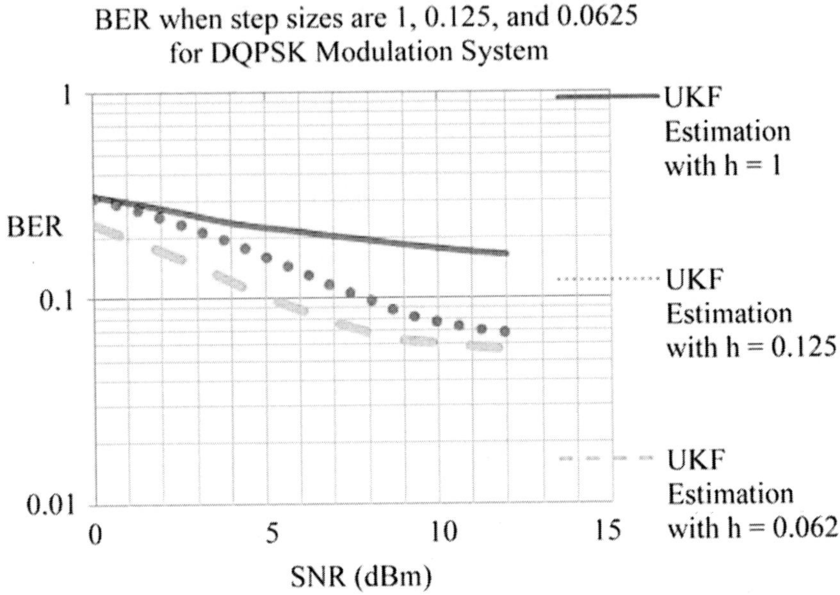

Figure 11. BER of the DQPSK demodulated signal when step sizes were chosen to be 1, 0.125, and 0.0625.

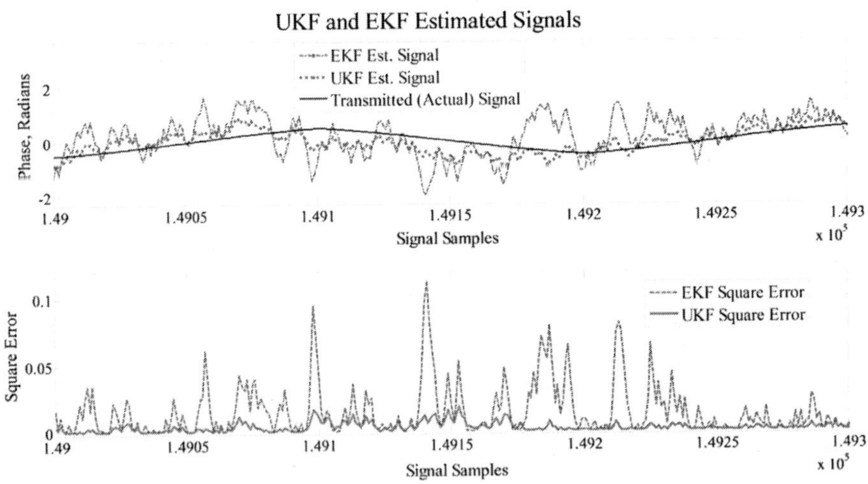

Figure 12. The estimated signals and associated square errors of the UKF and EKF for GFSK modulated signals.

Figures 7-10 simulated the performance when using both DQPSK and GFSK modulation in the simulated Bluetooth system model. The following figure represents performance when only DQPSK is used. **Figure 11** shows the simulation when the signals were UKF estimated signals for three values of the step size. In the first case, step size was chosen to be 1. The second and third step sizes were chosen to be 0.125 and 0.0625 to represent when 8 and 16 samples per symbol respectively.

When GFSK were the only modulation impeleneted in the Bluetooth system model as in the BDR, and simulation trials were run to compare the UKF and EKF performance, results were consistent in showing the superiroity of the UKF. **Figure 12** shows the comparison between the EKF and UKF estimated signals for both the phase signlas, and the associated SE. Results show clearly that UKF continues to produce bettr results than EKF.

For the GFSK only modulation system model, and in order to show that our proposed method is consistent over different types of randomly generated noises, we have run 10 simulation experiments for each SNR. Thus, for each SNR value, simulation has been run 10 times, using different values of noise seeds to generate different results in each trial. A total of 100 runs were executed, and when calculating the average of their MSEs, results showed clearly that UKF outperforms the EKF. The results, as they are shown in **Table 1** below, reveal the outcomes of these experiments.

Table 1. Average MSEs of the unfiltered, EKF, and UKF estimated signals.

SNR	Unfiltered MSE	EKF MSE	UKF MSE
1	1	0.411	0.079
5	0.750	0.285	0.062
10	0.376	0.092	0.029
15	0.101	0.032	0.011
20	0.025	0.010	0.006

7. CONCLUSION AND FUTURE WORK

In this paper, a novel IKF method has been proposed to optimize the performance of the non-coherent demodulation of DQPSK and GFSK. The results have demonstrated that this method is an effective when implemented in both modulation schemes, while preserving their simple structures. UKF has been proposed to be used in these digital modulation schemes, and has proven to be robust in handling the nonlinearity of both schemes and when operating in non-Gaussian noise environments. Utilizing UKF in DQPSK and GFSK has been validated using a Bluetooth communication system model, which has been created in MATLAB/SIMULINK. Results have shown the effectiveness and superiority of UKF over EKF.

This method can be expanded to be implemented in other modulation schemes. There is more work that can be done to enhance the idea of this research by adding the fading processes to the channels. Also, the analysis of the Bluetooth system model can be expanded to include the effect of frequency hopping and interferences, and IKF can be modified to handle these disturbances. Also, more researches using different types of sigma-points methods can be conducted for more optimum results.

REFERENCES

1. J. Proakis and M. Salehi, "Digital Communications," 5th Edition, McGraw-Hill Higher Education, New York, 2008.
2. M. Morelli, D. Benfatto, D. Lucano, M. Luise and U. Mengali, "Simple Non-Coherent Detectors for CPM Signals Transmitted over Rayleigh Flat-Fading Channels," Proceedings of the 4th IEEE Workshop on Signal Processing Advances in Wireless Communications, Rome, 15-18 June 2003, pp. 492-496.
3. W. P. Osborne and M. B. Luntz, "Coherent and Noncoherent Detection of CPFSK," IEEE Transactions on Communications, Vol. 22, No. 8, 1974, pp. 1023-1036.
4. Bluetooth Special Interest Group (SIG), "Specification of the Bluetooth System, Core Version 4.0," 2010. http://www.bluetooth.com
5. A. G. Soltanian and R. E. Van Dyck, "Performance of the Bluetooth System in Fading Dispersive Channels and Interference," Proceedings of the IEEE Global Telecommunications Conference, Vol. 6, 25-29 November 2001, pp. 3499-3503.
6. T. Aulin and C.-E. Sundbergm, "Continuous Phase Modulation-Part I: Full Response Signaling," IEEE Transactions on Communications, Vol. 29, No. 3, 1981, pp. 196-209.
7. L. Bin and P. Ho, "Data-Aided Linear Prediction Receiver for Coherent DPSK and CPM Transmitted over Rayleigh Flat-Fading Channels," IEEE Transactions on Vehicular Technology, Vol. 48, No. 4, 1999, pp. 1229-1236.

8. J. Anderson, T. Aulin and C. Sundbergm, "Digital Phase Modulation," Plenum Press, New York, 1986.

9. N. Ibrahim, L. Lampe and R. Schober, "Bluetooth Receiver Design Based on Laurent's Decomposition," IEEE Transactions on Vehicular Technology, Vol. 56, No. 4, 2007, pp. 1856-1862.

10. M. Nafie, A. Gatherer and A. Dabak, "Decision Feedback Equalization for Bluetooth Systems," Proceedings of the IEEE International Conference on Acoustics, Speech, and Signal Processing, Salt Lake City, May 2001, pp. 909- 912.

11. H. Leib, "Data-Aided Noncoherent Demodulation of DPSK," IEEE Transaction on Communications, Vol. 43, No. 234, 1995, pp. 722-725.

12. J. He, J. Cui, L. Yang and Z. Wang, "A Low-Complexity High-Performance Noncoherent Receiver for GFSK Signals," Proceedings of the IEEE International Symposium on Circuits and Systems, Seattle, 18-21 May 2008, pp. 1256-1259.

13. J. Gal, A. Campeanu and J. Nafornita, "Noncoherent Demodulation of Continuous Phase Modulated Signals Using Extended Kalman Filtering," Proceedings of the IEEE 12th International Conference on Optimization of Electrical and Electronic Equipment, Basov, 20-22 May 2010, pp. 724-727.

14. J. Gal, A. Campeanu and J. Nafornita, "Kalman Noncoherent Detection of CPFSK Signals," Proceedings of the 8th International Conference Communications, Bucharest, 10-12 June 2010, pp. 65-68.

15. O. Loffeld, "Demodulation of Noisy Phase or Frequency Modulated Signals with Kalman Filters," Proceedings of the IEEE Acoustics, Speech, and Signal Processing, Adelaide, 19-22 April 1994, pp. IV/177-IV/180.

16. R. E. Kalman, "A New Approach to Linear Filtering and Prediction Problems," Research Institute for Advanced Study, Baltimore, 1960.

17. M. S. Grewal and A. P. Andrews, "Kalman Filtering Theory and Practice Using MATLAB," 3rd Edition, John Wiley & Sons, Inc, New York, 2008.

18. J. Candy, "Bayesian Signal Processing Classical, Modern, and Particle Filtering Methods," John Wiley & Sons, Inc, New York, 2009.

19. D. Simon, "Optimal State Estimation, Kalman H∞, and Nonlinear Approaches," John Wiley & Sons, Inc, New York, 2009.

20. S. J. Julier and J. K. Uhlmann, "A New Extension of the Kalman Filter to Nonlinear Systems," The 11th International Symposium of Aerospace/Defense Sensing, Simulation and Controls, Multi Sensor Fusion, Tracking and Resource Management II, Orlando, 20-25 April 1997, pp. 182-193.

21. R. V. D. Merwe, "Sigma-Point Kalman Filters for Probabilistic Inference in Dynamic State-Space Models," Ph.D. Thesis, OGI School of Science & Engineering at Oregon Health & Science University, 2004.

22. S. J. Julier and J. K. Uhlmann, "Unscented Filtering and Nonlinear Estimation," Proceedings of the IEEE, Vol. 92, No. 3, 2004, pp. 401-422.

23. S. J. Julier, "The Scaled Unscented Transformation," Proceedings of the 2002 American Control Conference, Vol. 6, 2002, pp. 4555-4559.

24. S. J. Julier and J. K. Uhlmann, "Reduced Sigma Point Filters for the Propagation of Means and Covariances through Nonlinear Transformations," Proceedings of the 2002 American Control Conference, Vol. 2, 2002, pp. 887-892.

25. S. J. Julier and J. K. Uhlmann, "The Spherical Simplex Unscented Transformation," Proceedings of the American Control Conference, Denver, 4-6 June 2003, pp. 2430- 2434.

26. M. Ali and M. Zohdy, "Unscented Kalman Filtering for Continuous Phase Frequency Shift Keying Equalization", Proceedings of the International Conference on Information and Industrial Electronics, Chengdu, 14-15 January 2011, pp. 20-25.

27. T. S. Rappaport, "Wireless Communications Principles and Practice,"2nd Edition, Prentice Hall, Upper Saddle River, 2002.

28. C. E. Sundberg, "Continuous Phase Modulation," IEEE Communications Magazine, Vol. 24, No. 4, 1986, pp. 25- 38.

29. D. Labarre, E. Grivel, Y. Berthoumieu, E. Todini and M. Najim, "Consistent Estimation of Autoregressive Parameters from Noisy Observation Based on Two Interacting Kalman Filters," Signal Processing, Vol. 86, No. 10, 2006, pp. 2863-2876.

30. A. Jamoos, A. Abdo and H. Nour, "Estimation of OFDM Time-Varying Fading Channels Based on Two-CrossCoupled Kalman Filters," Novel Algorithms and Technique in Telecommunications, Automation and Industrial Electronics, 2008, pp. 287-292. http:// www. springerlink.com/content/978-1-4020-8736-3/#section=135616&page=1&locus=21

31. E. Punskaya, C. Andrieu, A. Doucet and W. J. Fitzgerald, "Particle Filtering for Demodulation in Fading Channels with Non-Gaussian Additive Noise," IEEE Transaction on Communications, Vol. 49, No. 4, 2001, pp. 579-582.

32. L. Binh, "MATLAB Simulink Simulation Platform for Photonic Transmission Systems," International Journal of Communications, Network and System Sciences, Vol. 2, 2009, pp. 91-168.

Chapter 2

AMA- and RWE- Based Adaptive Kalman Filter for Denoising Fiber Optic Gyroscope Drift Signal

Gongliu Yang [1,2], Yuanyuan Liu [1,2,*], Ming Li [1] and Shunguang Song [3]

[1]*School of Instrument Science and Opto-Electronics Engineering, Beihang University, Beijing 100191, China*

[2]*Inertial Technology Key Laboratory of National Defense Science and Technology, Beihang University, Beijing 100191, China*

[3]*Beijing Institute of Spacecraft System Engineering, Beijing 100094, China*

ABSTRACT

An improved double-factor adaptive Kalman filter called AMA-RWE-DFAKF is proposed to denoise fiber optic gyroscope (FOG) drift signal in both static and dynamic conditions. The first factor is Kalman gain updated by random weighting estimation (RWE) of the covariance matrix of innovation sequence at any time to ensure the lowest noise level of output, but the inertia of KF response increases in dynamic condition. To decrease the inertia, the second factor is the covariance matrix of predicted state vector adjusted by RWE only when discontinuities are detected by adaptive moving average (AMA).The AMA-RWE-DFAKF is applied for denoising FOG static and dynamic signals, its performance is compared with conventional KF (CKF), RWE-based adaptive KF with gain correction (RWE-AKFG), AMA- and RWE- based dual mode adaptive KF (AMA-RWE-DMAKF). Results of Allan variance on static signal and root mean square error (RMSE) on dynamic signal show that this proposed algorithm outperforms all the considered methods in denoising FOG signal.

Keywords: Adaptive Moving Average (AMA); Random Weighting Estimation (RWE); Fiber Optic Gyroscope (FOG); Kalman Filter (KF)

1. INTRODUCTION

As a kind of inertial sensor based on optical Sagnac effect, fiber optic gyroscope (FOG) has been widely used in inertial navigation system (INS) because of its significant advantages such as small size, low cost, no moving parts, long lifespan, and large dynamic range [1,2]. However, the accuracy of the FOG sensor is limited by drift errors due to internal device operations and external environment disturbances [3]. The FOG drift signal has two types of errors namely, deterministic errors and stochastic errors. Deterministic errors are bias, scale factor, and misalignment, which are relatively easier to be compensated by suitable calibration methods in laboratory environment. Stochastic errors, which are induced from the environmental temperature changes, electronic noises, and other electronic equipment interfaced with it [4,5], are difficult to be directly eliminated. As an alternative, stochastic models and denoising methods are two main techniques to restrain the FOG random errors reported in the literature.

Signal processing methods like low pass filter [6], wavelet transforms [7,8], and empirical mode decomposition (EMD) [9,10,11] are used to remove random errors for improving the performance of FOG measurement. These methods are successfully applied for denoising FOG static signal but fail to denoise FOG signal in highly dynamic conditions due to delay or complexity. Recently, time series analysis as a powerful tool is used to model the FOG random drift signal. Autoregressive (AR), moving average (MA), and autoregressive moving average (ARMA) have been developed in modeling stochastic models for FOG random errors in [12,13,14]. Combining these stochastic models, a conventional Kalman filter (CKF) is usually employed to remove the FOG random drift [13,14,15,16], where the process and measurement noises are pre-calculated by sampling lots of drift data. However, fixed noise variances are unsuitable in real applications which may lead to divergent problems. To improve the practicability and to avoid divergent effects, adaptive KF (AKF) methods have been investigated which are based on innovation-based adaptive estimation (IAE) or residual-based adaptive estimation (RAE) [17,18,19]. An AKF with double transitive factors is developed in [17,18], where the covariance matrix of predicted state vector is modified by an adaptive factor in stage one and the covariance matrix of measurement noise is modified by another adaptive factor in stage two. However, this method requires that innovation or residual vectors at each time point be in the identical type, spatial dimension and distribution, which is difficult to satisfy in a highly dynamic environment.

Random weighting estimation (RWE) is an advanced computational method in statistics. It does not require the prior knowledge of the distribution of position parameters, and the obtained estimation is unbiased. RWE has been established for estimation of the covariance matrix of observation vector and predicted state vector in [20,21,22]. In [23], Kalman gain based on RWE is updated using the covariance matrix of innovation sequence. This method called RWE-AKFG is applied to denoise the FOG signal under static and dynamic environments. However, the inertia of its response increases in denoising the FOG dynamic signal, which is not tolerable for real time dynamic applications. To solve the contradiction between the noise level of the output signal and the inertia of the KF response, adaptive moving average (AMA) is used to detect the discontinuities in signal [24]. AMA based dual mode KF (AMADMKF) in [25,26] is proposed to filter the FOG static and dynamic signals, where DMKF means the proper KF gain parameter or the ratio *Q/R* is switched at different conditions. However, it is difficult to predefine the proper ratio *Q/R* to adapt the FOG rotation rate changes [27]. Based on the above research, AMA-DWT-DMKF is used to denoise the FOG static signal, disturbance signal, and the change rate signal successfully in [28], but it is much more complex and has a higher computation cost. Considering the nonlinearity in system and measurement models, Narasimhappa *et al.* in [29,30,31] provide various adaptive filters based on unscented Kalman filter such as adaptive unscented Kalman filter (AUKF), adaptive square root unscented Kalman filter (ASRUKF) and adaptive sampling strong tracking scaled unscented Kalman filter (ASST-SUKF) for denoising the FOG signal. The performance of these algorithms is verified in static and dynamic conditions, whereas UKF is more complex than KF. Hence, the purpose of this paper is to develop an AMA- and RWE-based double-factor adaptive KF algorithm, named AMA-RWE-DFAKF, which can denoise the FOG static and dynamic signal. The first adaptive factor is Kalman gain, which is updated by using RWE of the covariance matrix of innovation sequence in any condition. The second adaptive factor is the covariance matrix of predicted state vector, which is revised based on RWE only when the discontinuities are detected by AMA. In fact, it has the same adaptive filter mechanism as [17], except that different estimations are used. Experimental results show that the proposed algorithm can satisfy the lowest noise level and the lowest inertia in denoising FOG static and dynamic signals.

The outline of this paper is as follows: Section 1 is the introduction; the concept of adaptive moving average is reviewed briefly in Section 2; Section 3 gives a description about the principle of random weighting estimation; the proposed AMA-RWE-DFAKF is provided in Section 4; experimental results and discussions about the proposed method applied in FOG static and dynamic signals are presented in Section 5, and Section 6 is the conclusion.

2. CONCEPT OF ADAPTIVE MOVING AVERAGE

The adaptive moving average (AMA) can be used to detect discontinuities in the signal by comparing the sample variance with a threshold value, where the length of moving average is adaptive to follow the rate of change in signal [24,25,26,28]. A q-point moving average filter can be expressed as

$$y(t) = \frac{1}{2q+1}\sum_{j=-q}^{q} x(t+j) \qquad q+1 \le t \le N-q \tag{1}$$

where $x(t)$ denotes the input data, $2q + 1$ is the moving average window size, N is the number of samples as one frame of the input data, and $y(t)$ is the filtered data. The noise can be further reduced by applying an iterative adaptive moving average filter shown in Figure 1. The algorithmic steps of this filter are explained as

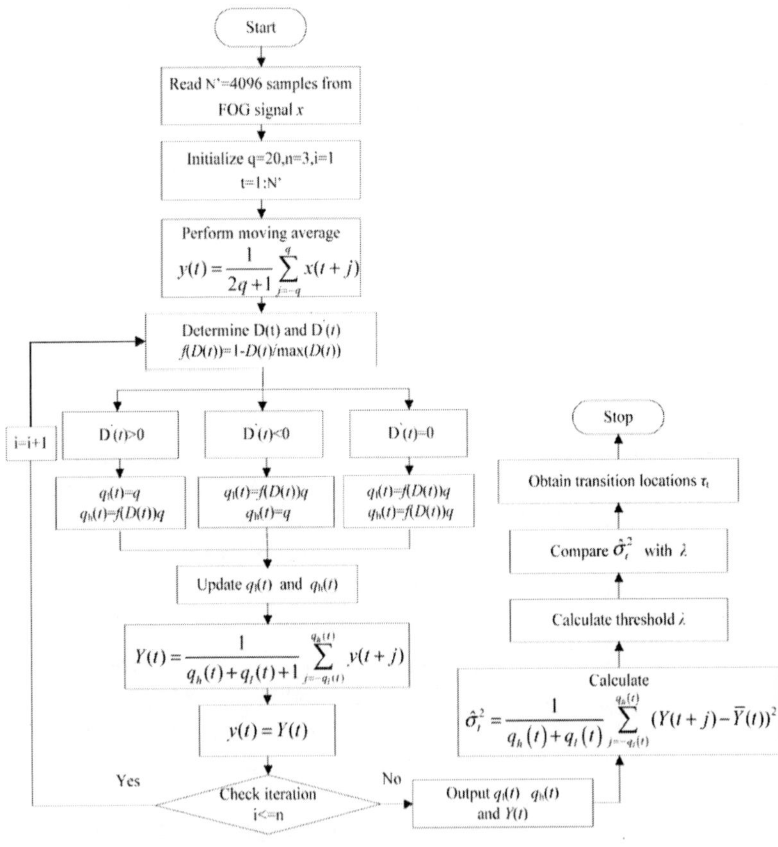

Figure 1. Procedure of the AMA algorithm.

Step 1: Calculate the absolute value of the differenced *y*(t)

$$D(t) = \left| y(t+q) - y(t-q) \right| \tag{2}$$

Step 2: Calculate the rate of change of *D*(t)

$$D^{'}(t) = D(t+1) - D(t) \tag{3}$$

Step 3: Calculate the adaptive filtered data *Y*(t)

$$Y(t) = \frac{1}{q_h(t) + q_l(t) + 1} \sum_{j=-q_l(t)}^{q_h(t)} y(t+j) \qquad q + q_l + 1 \le t \le N - q - q_h \tag{4}$$

where

$$q_h(t) = \begin{cases} q & if\ D^{'}(t) < 0 \\ f(D(t))q & if\ D^{'}(t) \ge 0 \end{cases} \tag{5}$$

$$q_l(t) = \begin{cases} q & if\ D^{'}(t) > 0 \\ f(D(t))q & if\ D^{'}(t) \le 0 \end{cases} \tag{6}$$

$$f(D(t)) = 1 - \frac{D(t)}{\max\left(D(t)\right)} \tag{7}$$

Step 4: Repeat the iteration on the filtered data *Y*(*t*) from step 1 to 3 until the maximum iterations is met.

The final filtered data *Y*(*t*) is obtained by *n* iteration in our work, *i.e.*, *n* = 3. And the transition locations are detected by comparing the sample variance of *Y*(*t*) with a threshold λ. The sample variance is calculated in the window size as follows

$$\hat{\sigma}_t^2 = \frac{1}{q_h(t) + q_l(t)} \sum_{j=-q_l(t)}^{q_h(t)} (Y(t+j) - \bar{Y}(t))^2 \tag{8}$$

The threshold λ is 95% upper tail of exponential distribution, where the expected value of this distribution is the mean value of the above calculated sample variances in the current frame. The transition location τ_t is defined as

$$\tau_t = (t \mid \hat{\sigma}_t^2 > \lambda, t = 3q+1, 3q+2, \cdots N-3q-1) \tag{9}$$

3. PRINCIPLE OF RANDOM WEIGHTING ESTIMATION

Suppose that $X_1, X_2, \ldots, X_n$ are the independent and identically distributed random variables with common distribution function $F(x)$. Let $x_1, x_2, \ldots, x_n$ be the sample realizations. The corresponding empirical distribution function $F_n(x)$ can be expressed as

$$F_n(x) = \frac{1}{n}\sum_{i=1}^{n} I_{(X_i \le x)} \tag{10}$$

The random weighting estimation [21,22,23] of $F_n(x)$ is defined as

$$F_n^*(x) = \sum_{i=1}^{n} V_i I_{(X_i \le x)} \tag{11}$$

where $I_{(X_i \le x)}$ is the indicator function represented as

$$I_{(X_i \le x)} = \begin{cases} 1 & X_i \le x \\ 0 & X_i > x \end{cases} \tag{12}$$

and random vector $[V_1, V_2, \ldots, V_n]$ subjects to Dirichlet distribution $D(1, 1, \ldots, 1)$, that is, $\sum_{i=1}^{n} V_i = 1$. A joint density function of the random vector is defined as

$$f(V_1, V_2, \ldots, V_n) = (n-1)! \tag{13}$$

where $[V_1, V_2, \ldots, V_n] \in D_n$, and $D_{n-1}=\{[V_1, V_2, \ldots, V_{n-1}]: V_i \ge 0$ $(I = 1, 2, \ldots, n-1)$, $\sum_{i=1}^{n-1} V_i \le 1\}$.

Calculate the mean of X_i as follows

$$\hat{E}_X = \frac{1}{n}\sum_{i=1}^{n} X_i \tag{14}$$

Then, the random weighting estimation of E^X is

$$\hat{E}_X^* = \sum_{i=1}^{n} v_i X_i \tag{15}$$

Calculate the variance of X_i as follows

$$\hat{\Sigma}_X = \frac{1}{n}\sum_{i=1}^{n}[X_i - E(X)][X_i - E(X)]^T \tag{16}$$

Then, the random weighting estimation of $\hat{\Sigma}_X$ is

$$\hat{\Sigma}_X^* = \sum_{i=1}^{n} v_i [X_i - E(X)][X_i - E(X)]^T \tag{17}$$

4. ADAPTIVE KALMAN FILTERING

4.1. Conventional Kalman Filter

As an efficient and recursive estimator, Kalman filter has been widely used for eliminating the random noise of FOG sensor [18,25,28,31]. It is a set of mathematical equations to estimate the state of system and minimize the mean squared error of residuals using the prior knowledge about dynamic process and measurement models, in addition to the process and measurement noise. Let the linear dynamic system and measurement equations be given by

$$X_k = \Phi_{k,k-1} X_{k-1} + \Gamma_{k-1} W_{k-1} \tag{18}$$

$$Z_k = H_k X_k + V_k \tag{19}$$

where X_k is the state vector at epoch k, Z_k is measurement vector, $\Phi_{k,k-1}$ is the state transition matrix, Γ_{k-1} is the system noise driving matrix, and H_k denotes the measurement matrix. W_k is the state noise with covariance matrix ∑Wk, V_k is the measurement noise with covariance matrix ∑Vk. W_k and V_k are assumed to be discrete white Gaussian noise with zero mean, known distributions and uncorrelated to each other, satisfying

$$\begin{cases} E[W_k]=0 \quad E\left[W_k W_j^T\right]=Q_k\delta_{kj} \\ E[V_k]=0 \quad E\left[V_k V_j^T\right]=R_k\delta_{kj} \\ E\left[W_k V_j^T\right]=0 \end{cases} \tag{20}$$

$$\delta_{kj}=\begin{cases} 1 & k=j \\ 0 & k\neq j \end{cases} \tag{21}$$

where Q_k is process noise covariance matrix, R_k is measurement noise variance matrix, and δ_{kj} is the Kronecker δ function. The CKF algorithm is described by the following equations

$$\hat{X}_{k,k-1}=\Phi_{k,k-1}\hat{X}_{k-1} \tag{22}$$

$$P_{k,k-1}=\Phi_{k,k-1}P_{k-1}\Phi_{k,k-1}^T+\Gamma_{k-1}Q_{k-1}\Gamma_{k-1}^T \tag{23}$$

$$K_k=P_{k,k-1}H_k^T(H_kP_{k,k-1}H_k^T+R_k)^{-1} \tag{24}$$

$$\hat{X}_k=\hat{X}_{k,k-1}+K_k(Z_k-H_k\hat{X}_{k,k-1}) \tag{25}$$

$$P_k=(I-K_kH_k)P_{k,k-1} \tag{26}$$

4.2. Random Weighting Estimation for Kalman Gain

The predicted state vector is

$$\hat{X}_{k,k-1} = \Phi_{k,k-1}\hat{X}_{k-1} \tag{27}$$

Accordingly, the innovation vector may be written as

$$V_{\hat{X}_{k,k-1}} = Z_k - H_k\hat{X}_{k,k-1} \tag{28}$$

The variance of $V_{\hat{X}_{k,k-1}}$ is expressed as

$$\hat{\Sigma}_{V_{\hat{X}_{k,k-1}}} = \frac{1}{N}\sum_{i=1}^{N}[V_{\hat{X}_{k-i,k-i-1}} - E(V_{\hat{X}_{k-i,k-i-1}})][V_{\hat{X}_{k-i,k-i-1}} - E(V_{\hat{X}_{k-i,k-i-1}})]^T \tag{29}$$

where $E(V_{\hat{X}_{k-i,k-i-1}}) = 0.$ Accordingly, the random weighting estimation of $\hat{\Sigma}_{V_{\hat{X}_{k,k-1}}}$ is

$$\hat{\Sigma}^*_{V_{\hat{X}_{k,k-1}}} = \sum_{i=1}^{N} v_i[V_{\hat{X}_{k-i,k-i-1}} - E(V_{\hat{X}_{k-i,k-i-1}})][V_{\hat{X}_{k-i,k-i-1}} - E(V_{\hat{X}_{k-i,k-i-1}})]^T = \sum_{i=1}^{N} v_i V_{\hat{X}_{k-i,k-i-1}} V^T_{\hat{X}_{k-i,k-i-1}} \tag{30}$$

The variance of $V_{\hat{X}_{k,k-1}}$ can be rewritten as

$$\begin{aligned}\Sigma_{V_{\hat{X}_{k,k-1}}} &= E\left\{[V_{\hat{X}_{k,k-1}} - E(V_{\hat{X}_{k,k-1}})][V_{\hat{X}_{k,k-1}} - E(V_{\hat{X}_{k,k-1}})]^T\right\} \\ &= E\left\{[Z_k - E(Z_k)][Z_k - E(Z_k)]^T\right\} + E\left\{[E(H_k\hat{X}_{k,k-1}) - H_k\hat{X}_{k,k-1}][E(H_k\hat{X}_{k,k-1}) - H_k\hat{X}_{k,k-1}]^T\right\} \\ &= \Sigma_{Z_k} + H_k(\Sigma_{\hat{X}_{k,k-1}})H_k^T = H_k P_{k,k-1} H_k^T + R_k\end{aligned} \tag{31}$$

By substituting Equation (31) into (24) and considering (30), the modified gain K_k can be described as

$$K_k^* = P_{k,k-1}H_k^T(\hat{\Sigma}^*_{V_{\hat{X}_{k,k-1}}})^{-1} \tag{32}$$

4.3. Random Weighting Estimation for Covariance Matrix of Predicted State Vector

In KF prediction stage, the covariance matrix of predicted state vector is

$$P_{k,k-1} = \Phi_{k,k-1} P_{k-1} \Phi_{k,k-1}^T + \Gamma_{k-1} Q_{k-1} \Gamma_{k-1}^T \tag{33}$$

That is

$$\Sigma_{\hat{X}_{k,k-1}} = \Phi_{k,k-1} \Sigma_{\hat{X}_{k-1}} \Phi_{k,k-1}^T + \Gamma_{k-1} \Sigma_{W_{k-1}} \Gamma_{k-1}^T \tag{34}$$

By Equation (34), the covariance matrix of W_{k-1} can be written as

$$\Gamma_{k-1} \Sigma_{W_{k-1}} \Gamma_{k-1}^T = \Sigma_{\hat{X}_{k,k-1}} - \Phi_{k,k-1} \Sigma_{\hat{X}_{k-1}} \Phi_{k,k-1}^T = \Sigma_{\hat{X}_{k,k-1} - \hat{X}_k} + \Sigma_{\hat{X}_k} - \Phi_{k,k-1} \Sigma_{\hat{X}_{k-1}} \Phi_{k,k-1}^T \tag{35}$$

Take the average of $\Sigma_{W_{k-i-1}}$ as the estimation of $\Sigma_{W_{k-1}}$ yields

$$\Gamma_{k-1} \hat{\Sigma}_{W_{k-1}} \Gamma_{k-1}^T = \frac{1}{N} \sum_{i=1}^{N} \Gamma_{k-i-1} \Sigma_{W_{k-i-1}} \Gamma_{k-i-1}^T \tag{36}$$

Considering Equation (35), the random weighting estimation of $\Sigma_{W_{k-1}}$ is

$$\begin{aligned} \Gamma_{k-1} \hat{\Sigma}_{W_{k-1}}^* \Gamma_{k-1}^T &= \sum_{i=1}^{N} v_i \Gamma_{k-i-1} \Sigma_{W_{k-i-1}} \Gamma_{k-i-1}^T \\ &= \sum_{i=1}^{N} v_i \left(\Sigma_{\hat{X}_{k-i,k-i-1} - \hat{X}_{k-i}} + \Sigma_{\hat{X}_{k-i}} - \Phi_{k-i,k-i-1} \Sigma_{\hat{X}_{k-i-1}} \Phi_{k-i,k-i-1}^T \right) \\ &= \hat{\Sigma}_{\hat{X}_{k,k-1} - \hat{X}_k}^* + \hat{\Sigma}_{\hat{X}_k}^* - \Phi_{k,k-1} \hat{\Sigma}_{\hat{X}_{k-1}}^* \Phi_{k,k-1}^T \end{aligned} \tag{37}$$

Accordingly, we have

$$\Gamma_{k-1} \hat{Q}_{k-1}^* \Gamma_{k-1}^T = \hat{\Sigma}_{\hat{X}_{k,k-1} - \hat{X}_k}^* + \hat{\Sigma}_{\hat{X}_k}^* - \Phi_{k,k-1} \hat{\Sigma}_{\hat{X}_{k-1}}^* \Phi_{k,k-1}^T \tag{38}$$

Take the average of ∑X^k-i,k-i-1 as the estimation of ∑X^k,k−1 yields

$$\hat{\Sigma}_{\hat{X}_{k,k-1}} = \frac{1}{N}\sum_{i=1}^{N}\Sigma_{\hat{X}_{k-i,k-i-1}} \tag{39}$$

The random weighting estimation of ∑X^k,k−1 is

$$\hat{\Sigma}^{*}_{\hat{X}_{k,k-1}} = \sum_{i=1}^{N} v_i \Sigma_{\hat{X}_{k-i,k-i-1}} \tag{40}$$

Substituting Equation (34) into (40), we have

∑^∗X^k,k−1=∑i=1Nvi(Φk-i,k-i-1∑X^k-i-1ΦTk-i,k-i-1+Γk-i-1Qk-i-1ΓTk-i-1)=Φk,k-1∑^∗X^k-1ΦTk,k-1+Γk-1Q^∗k−1ΓTk-1

$$\hat{\Sigma}^{*}_{\hat{X}_{k,k-1}} = \sum_{i=1}^{N} v_i(\Phi_{k-i,k-i-1}\Sigma_{\hat{X}_{k-i-1}}\Phi^{T}_{k-i,k-i-1} + \Gamma_{k-i-1}Q_{k-i-1}\Gamma^{T}_{k-i-1}) = \Phi_{k,k-1}\hat{\Sigma}^{*}_{\hat{X}_{k-1}}\Phi^{T}_{k,k-1} + \Gamma_{k-1}\hat{Q}^{*}_{k-1}\Gamma^{T}_{k-1} \tag{41}$$

Substituting Equation (38) into (41), we have

$$\begin{aligned}\hat{\Sigma}^{*}_{\hat{X}_{k,k-1}} &= \Phi_{k,k-1}\hat{\Sigma}^{*}_{\hat{X}_{k-1}}\Phi^{T}_{k,k-1} + \hat{\Sigma}^{*}_{\hat{X}_{k,k-1}-\hat{X}_k} \\ &+ \hat{\Sigma}^{*}_{\hat{X}_{k,k-1}-\hat{X}_k} + \hat{\Sigma}^{*}_{\hat{X}_k} - \Phi_{k,k-1}\hat{\Sigma}^{*}_{\hat{X}_{k-1}}\Phi^{T}_{k,k-1} = \hat{\Sigma}^{*}_{\hat{X}_{k,k-1}-\hat{X}_k} + \hat{\Sigma}^{*}_{\hat{X}_k}\end{aligned} \tag{42}$$

The error question of predicted state vector can be defined as

$$V_{\bar{X}_k} = \hat{X}_{k,k-1} - \hat{X}_k \tag{43}$$

The random weighting estimation of ∑VX---k is

$$\hat{\Sigma}^{*}_{V_{\bar{X}_k}} = \sum_{i=1}^{N} v_i[V_{\bar{X}_{k-i}} - E(V_{\bar{X}_{k-i}})][V_{\bar{X}_{k-i}} - E(V_{\bar{X}_{k-i}})]^{T} \tag{44}$$

The random weighting estimation of ∑X^k is

$$\hat{\Sigma}^{*}_{\hat{X}_k} = \sum_{i=1}^{N} v_i[\hat{X}_{k-i} - E(\hat{X}_{k-i})][\hat{X}_{k-i} - E(\hat{X}_{k-i})]^{T} \tag{45}$$

Considering Equations (42)–(45), the covariance matrix of predicted state vector $\hat{\Sigma}^*_{\hat{X}_{k,k-1}}$ can be updated by RWE as follows

$$\hat{\Sigma}^*_{\hat{X}_{k,k-1}} = \hat{\Sigma}^*_{\hat{X}_{k,k-1}-\hat{X}_k} + \hat{\Sigma}^*_{\hat{X}_k} = \hat{\Sigma}^*_{V_{\bar{X}_k}} + \hat{\Sigma}^*_{\hat{X}_k}$$
$$= \sum_{i=1}^{N} v_i [V_{\bar{X}_{k-i}} - E(V_{\bar{X}_{k-i}})][V_{\bar{X}_{k-i}} - E(V_{\bar{X}_{k-i}})]^T + \sum_{i=1}^{N} v_i [\hat{X}_{k-i} - E(\hat{X}_{k-i})][\hat{X}_{k-i} - E(\hat{X}_{k-i})]^T \tag{46}$$

KF parameters Q_k, R_k, and P_k impact not only on the noise level of output signal, but also on the inertia of KF response [27]. The inertia increases with the value of R_k while the noise level decreases. Conversely, the inertia decreases with the increase of value of Q_k while the noise level increases. In the dynamic case, these fixed values are critical in KF denoising scheme due to the noise characteristic of the FOG is more complex and time varying. To adjust KF parameters for the real-time applications, a double-factor adaptive KF combined AMA and RWE as shown in Figure 2, called AMA-RWE-DFAKF, is proposed for denoising the FOG signal. The Kalman gain is updated as Equation (32) in any region and the covariance matrix of predicted state vector is modified as Equation (46) only in transition region, which in fact has the same adaptive filter mechanism as [17], except that different methods are used to estimate these adaptive parameters.

5. EXPERIMENTAL RESULTS AND DISCUSSIONS

In this experiment, we test the filtering on real FOG signal under both static and dynamic environments. The experimental setup is shown in Figure 3. The setup has a single-axis FOG, three-axis turntable, power supply for FOG, and data processing computer. In the static condition, FOG is in zero rotation at room temperature, whereas in the dynamic condition, FOG is mounted on the three-axis turntable with different rotation rates.

5.1. Static Test Analysis

Under room temperature, the FOG static data is recorded for 3 h with a sampling frequency of 100 Hz as shown in Figure 4. In KF, an AR (2) model is established as the system state equation using the first 6000 samples. Here, initial state and error covariance matrix of the state are assumed to X^0=[0, 0]T and P_0 = *diag*([1, 1]). H = [1, 0] is the observation matrix. And the measurement (*R*) and process (*Q*) noise covariance matrix are initialized to 0.01 and 0.0001, respectively. All these considered algorithms have the same parameters. Details on RWE-AKFG and AMA-DMKF are fully available

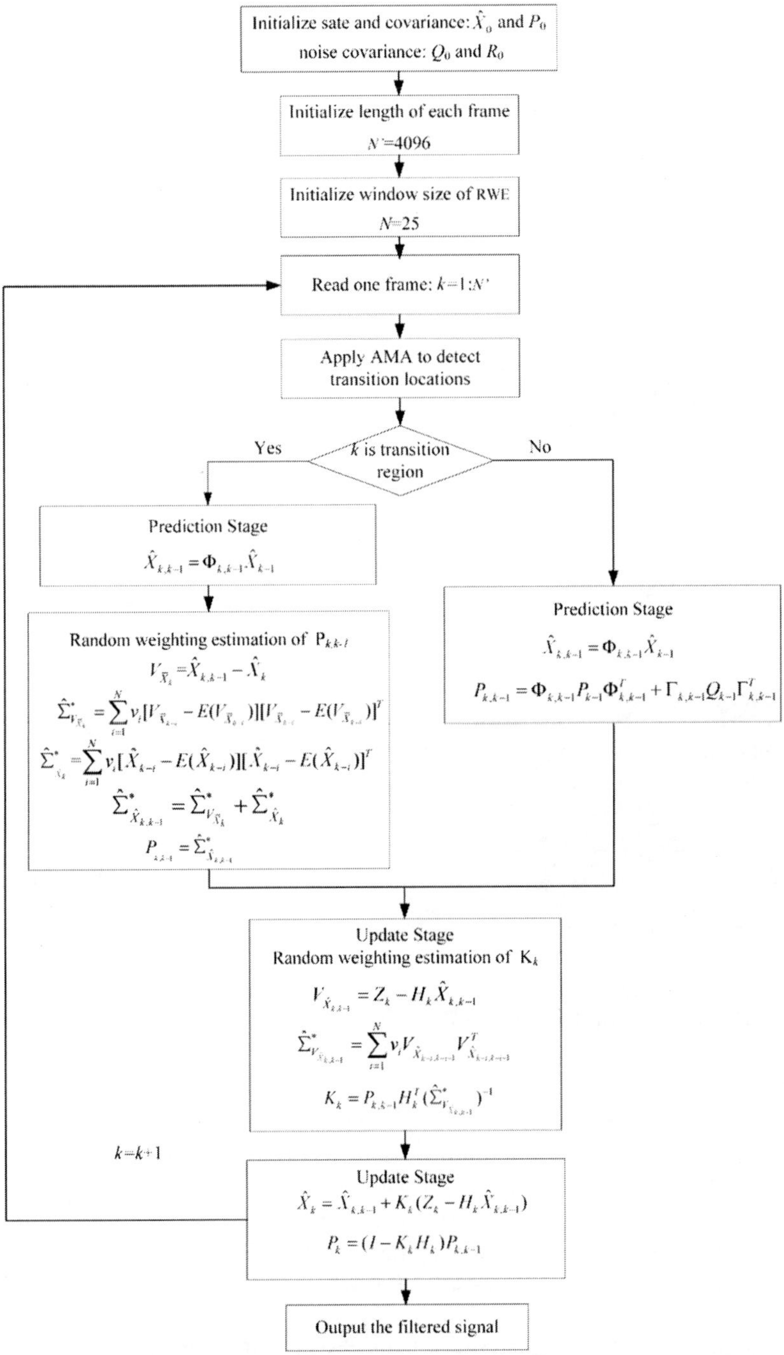

Figure 2. Operation of the proposed AMA-RWE-DFAKF algorithm.

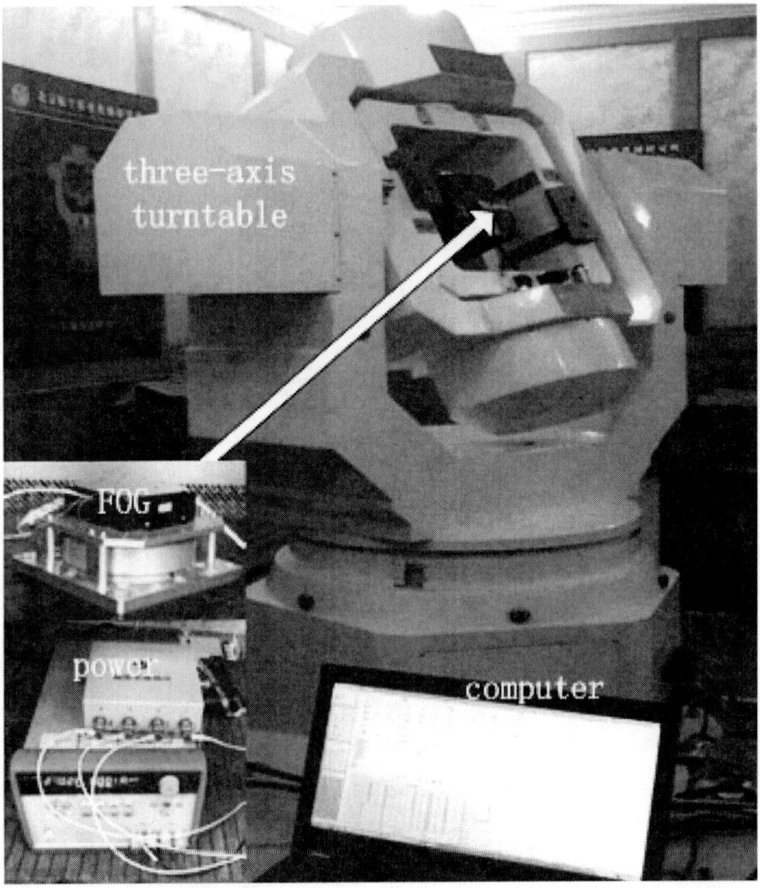

Figure 3. Experimental setup of FOG.

in [23,24,25,26]. In this paper, *Q/R* is chosen as 0.01 ($k1$) and 0.1 ($k2$) for non-discontinuity region and discontinuity region, respectively. As discussed in [27], the value of *R* is fixed. AMA-RWE-DMAKF is that Kalman gain is updated by the covariance matrix of innovation sequence using random weighting method, which is a combination of RWE-AKFG and AMA-DMKF. For AMA-RWE-DFAKF and AMA-RWE-DMAKF algorithms, we considered 4096 samples as one frame. However, for RWE-AKFG and CKF we denoised the signal sample by sample. There is no discontinuity location detected by AMA because of collecting FOG signal in motionless environment.

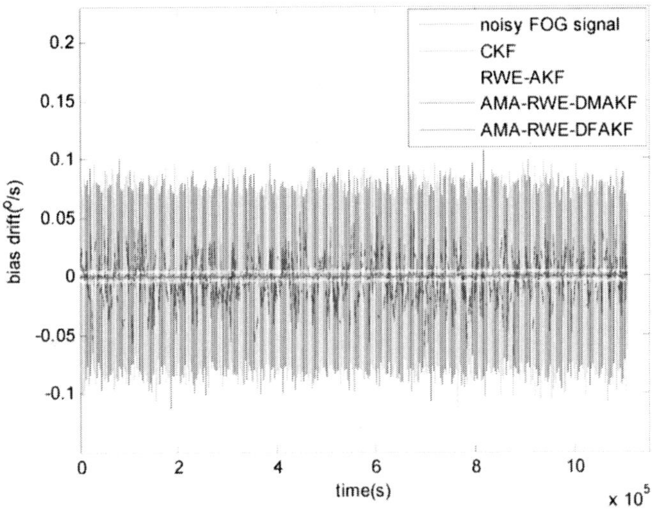

Figure 4. Denoising results of FOG signal in static condition.

Allan variance is a popular technique to quantify and identify the random noise like quantization noise (Q), angle random walk (N), bias instability (B), rate random walk (K) and rate ramp (R) before and after denoising the FOG static drift signal [11,26]. In Allan variance curve, different slope corresponds to different random noise. To compare the denoising performance, Allan variance curves are plotted in Figure 5. It can be seen that the curves after filtering by these four methods decline in some degree and the curve from the proposed method is the lowest. In Figure 5, slopes of −1/2 and 0 indicate the present of angle random walk and bias instability in this FOG signal. The random error values are tabulated in Table 1. It is seen that the angle random walk and bias instability are reduced by 100 times as compared to the original value. Moreover, AMA-RWE-DFAKF, AMA-RWE-DMAKF, and RWE-AKFG algorithms give a competitive performance in denoising FOG static signal, and these results have a clear advantage as compared with CKF.

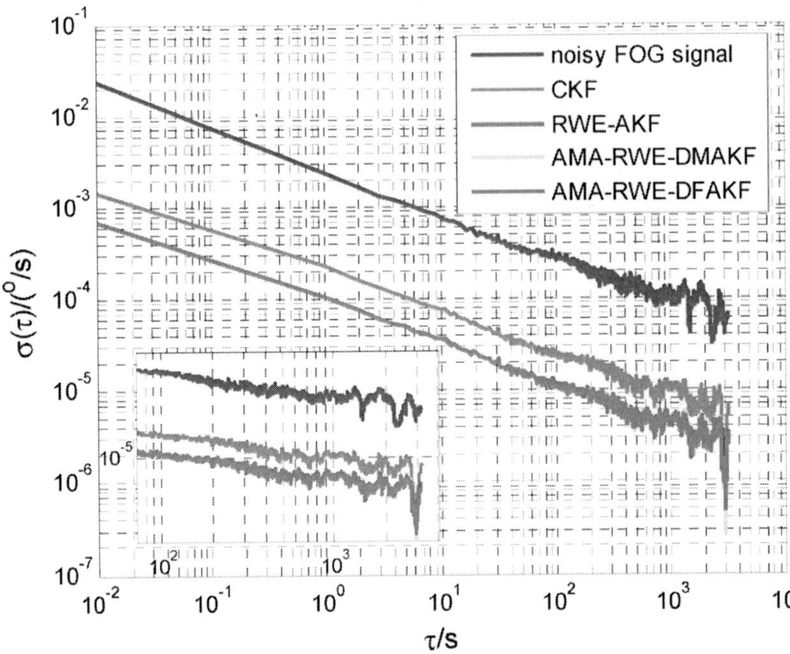

Figure 5. Allan variance analysis of FOG signal in static condition.

Table 1. Allan variance analysis results of FOG signal in static condition.

Methods	Q (μrad)	N ($o/\sqrt{h}$)	B (o/h)	K ($o/h^{3/2}$)	R (o/h^2)
Input	-	2.829×10^{-3}	4.065×10^{-4}	-	-
CKF	-	1.446×10^{-4}	4.969×10^{-6}	-	-
RWE-AKFG	-	6.702×10^{-5}	2.398×10^{-6}	-	-
AMA-RWE-DMAKF	-	6.711×10^{-5}	2.282×10^{-6}	-	-
AMA-RWE-DFAKF	-	6.705×10^{-5}	2.264×10^{-6}	-	-

5.2. Dynamic Test Analysis

For acquiring the dynamic FOG data, the three-axis turntable is used to generate series of reference angular rate in the dynamic ranges, *i.e.*, ±10, ±20, and ±50°/s. In this study, the signal data is recorded for 1 h at room temperature with sampling frequency of 100 Hz shown in Figure 6 through a clockwise and counter-clockwise turntable rotation. The rotation rate is decreased in a stepwise manner starting from +50°/s and varying between 0°/s and 50°/s. An AR (2) model is established as the system state equation using the first 6000 differentiated samples. Here, the initial state and error covariance matrix of the state are assumed to X^0 = $[0, 0, 0]^T$ and P_0 =

diag([1, 1, 1]), respectively. The observation matrix becomes H = [1, 0, 0]. The initial values of *R* and *Q* are fixed to 0.01 and 0.0001 as mentioned above. The ratio between *Q* and *R* is switched from two states 0.01 and 0.1 for AMA-RWE-DMAKF.

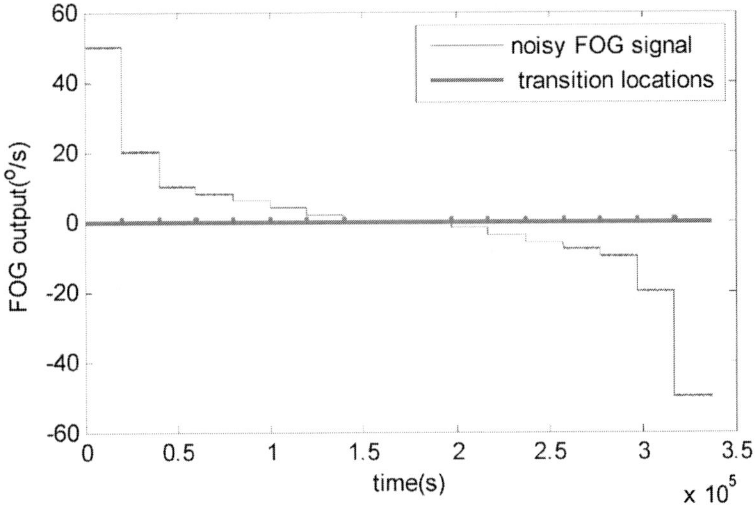

Figure 6. Detected discontinuities using AMA.

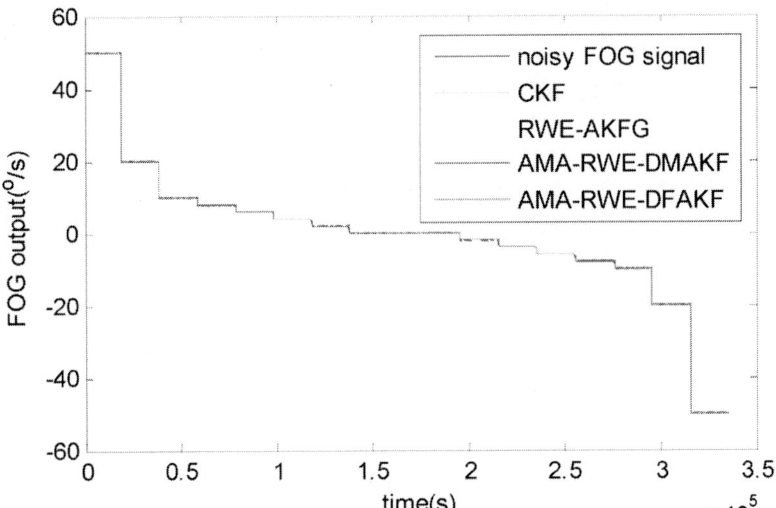

Figure 7. Denoising results of FOG dynamic signal.

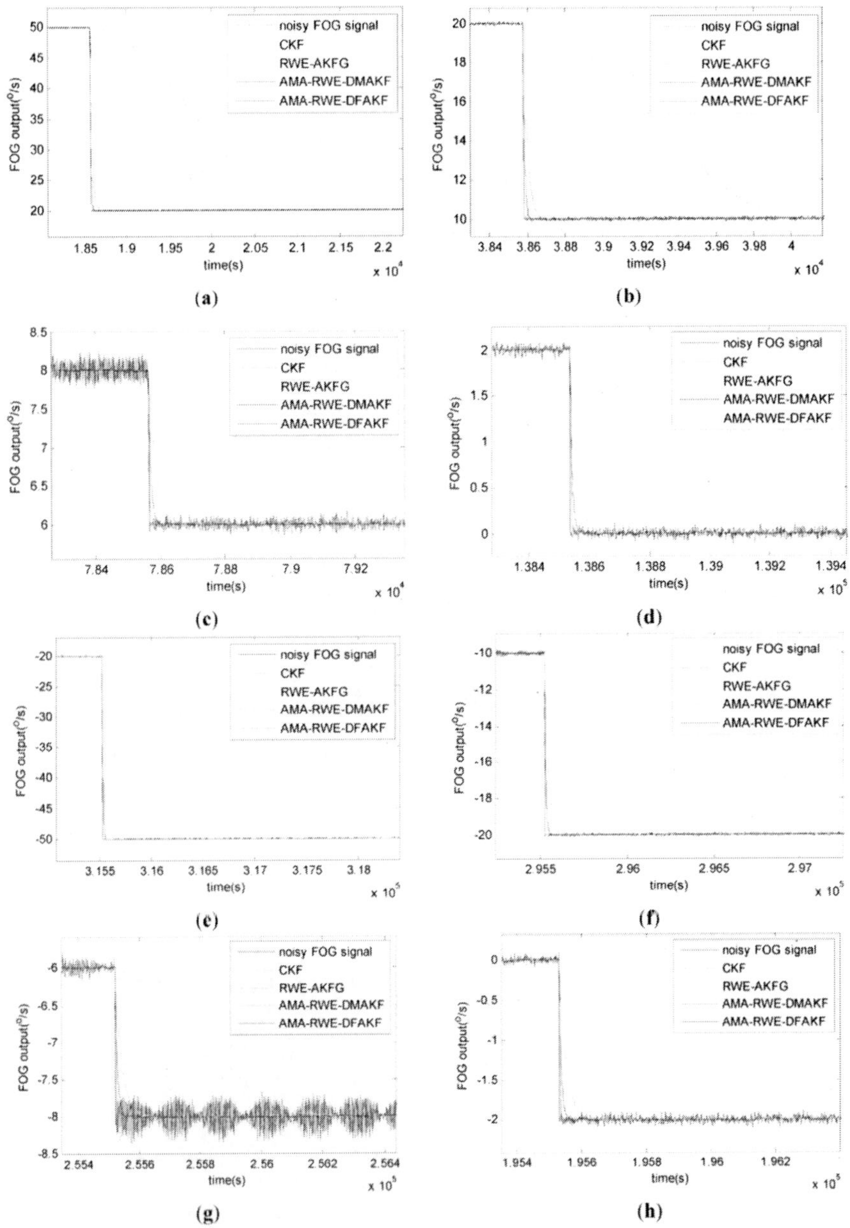

Figure 8. Comparison of denoising results for FOG dynamic signal at different rotations. (**a**) Rotation rate from 50 to 20°/s; (**b**) Rotation rate from 20 to 10°/s; (**c**) Rotation rate from 8 to 6°/s; (**d**) Rotation rate from 2 to 0°/s; (**e**) Rotation rate from −20 to −50°/s; (**f**) Rotation rate from −10 to −20°/s; (**g**) Rotation rate from −6 to −8°/s; (**h**) Rotation rate from 0 to −2°/s.

As already mentioned, the signal is divided into frames with $N' = 4096$ samples. For each frame, AMA is used to detect discontinuity locations shown in Figure 6. From these results, the effectiveness of AMA is proved. These four algorithms are applied to denoise FOG dynamic signal with the same initial parameters chosen as in the static condition, *i.e.*, the measurement and process noise covariance matrix. Although RWE-AKFG and AMA-RWE-DMAKF give quite competitive results with AMA-RWE-DFAKF in static condition, but these fail to denoise signal in dynamic condition.

The denoising results for all samples are shown in Figure 7. To obtain a clear visuality, we have plotted only a portion of the signal in Figure 8 where we can compare the inertia of these four algorithms in the different transition. From Figure 8, we can see there exists some delay for every filtering in following trend of the signal. It is not acceptable for RWE-AKF due to the increase of the value of R while Q is fixed. Thus AMA-RWE-DMAKF with two Kalman gain $k1$ and $k2$ is developed to decrease the inertial, but it is difficult to pre-design these parameters. However, the proposed AMA-RWE-DFAKF algorithm can satisfy the lowest noise level and the lowest inertial by only adjusting the two parameters adaptively as described in Figure 2. Figure 7 and Figure 8 indicate that the AMA-RWE-DFAKF denosies the FOG dynamic signal better than all other algorithms.

To further verify the effectiveness of AMA-RWE-DFAKF in denoising the FOG dynamic signal, we apply the same procedures on another FOG made in different company. The rotation rate of the table is increased by the alternatively positive and negative variations from 0°/s to 200°/s. Step signals data are collected for 1 h with sampling frequency of 200 Hz shown in Figure 9 and Figure 10. For comparing denoising results clearly, the zoomed figures of denoised signal in different rotation rates are plotted in Figure 11. The novel AMA-RWE-DFAKF has the minimum lag at transition times, which can satisfy both conditions of the lowest noise level and the lowest inertia.

Mean square error (MSE), root mean square error (RMSE), or signal-to-noise power ratio (SNR) [11,31] are generally employed to compare the performance of different denoising methods before and after denoising the FOG dynamic drift signal. The RMSE is defined as follows

$$RMSE = \sqrt{\frac{1}{N}\sum_{t=1}^{N}(x(t)-\hat{x}(t))^2} \tag{47}$$

where $\hat{x}(t)$ is the denoised signal, $x(t)$ is the actual signal and N is the number of signal.

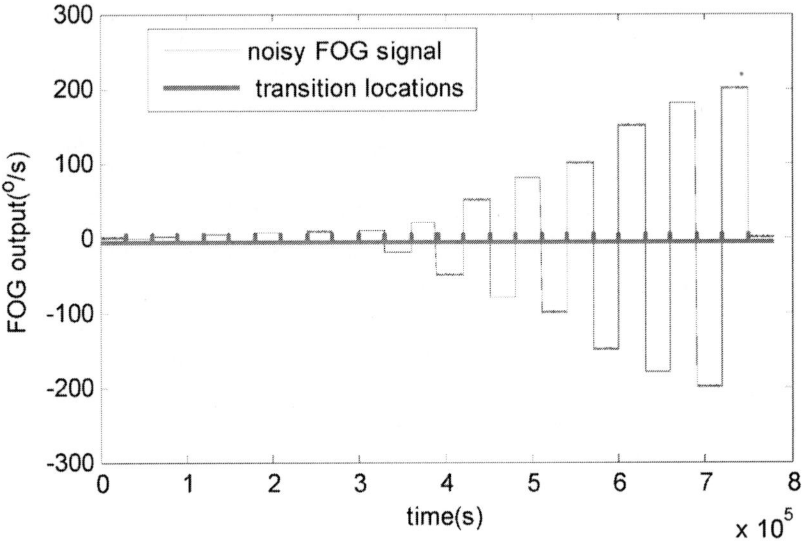

Figure 9. Detected discontinuities using AMA.

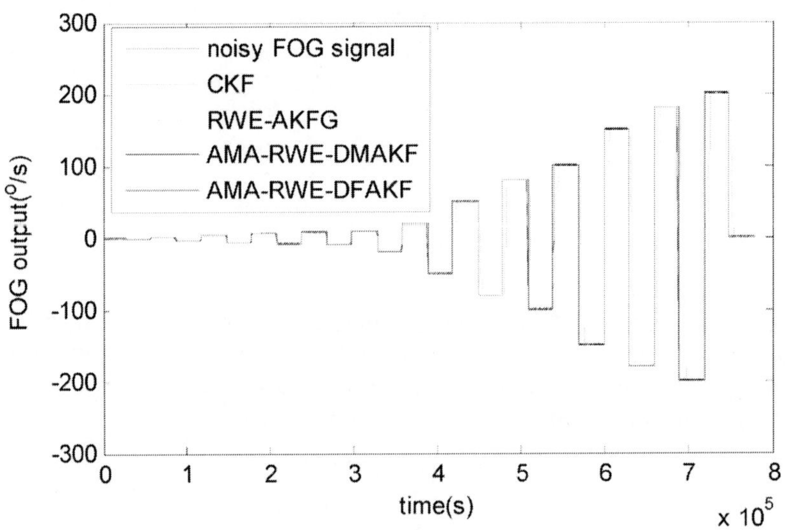

Figure 10. Denoising results of FOG dynamic signal.

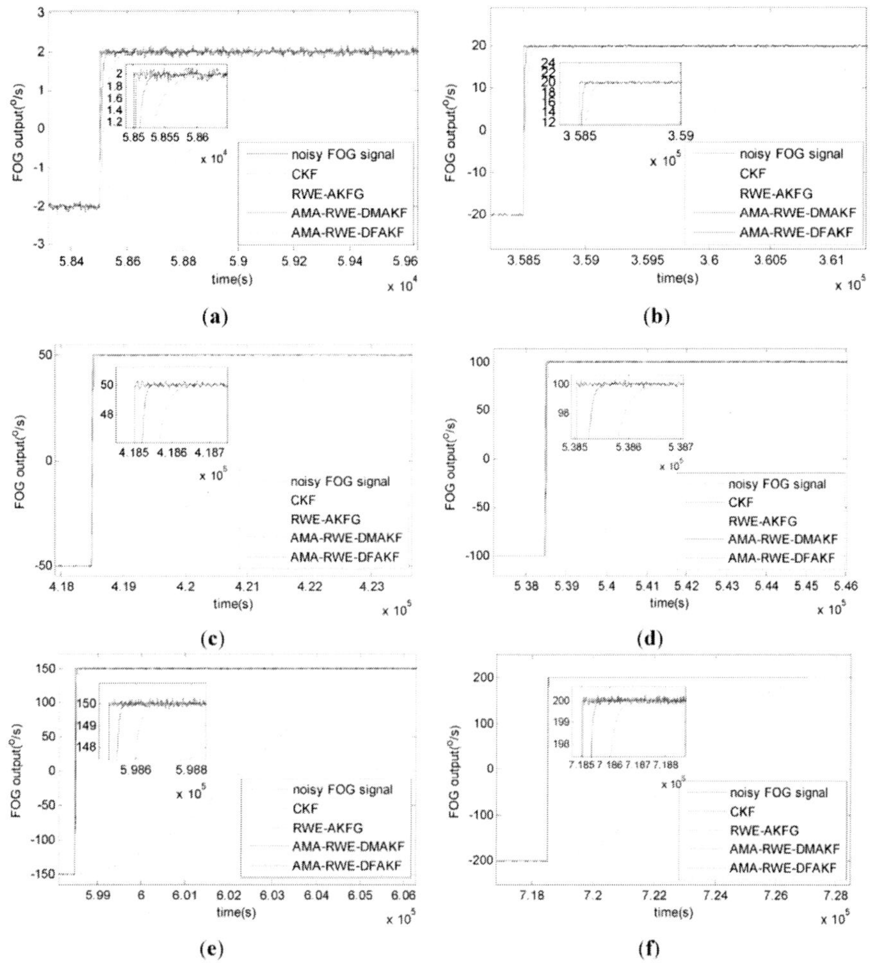

Figure 11. Comparison of denoising results for FOG dynamic signal at different rotations. (**a**) Rotation rate from −2 to 2°/s; (**b**) Rotation rate from −20 to 20°/s; (**c**) Rotation rate from −50 to 50°/s; (**d**) Rotation rate from −100 to 100°/s; (**e**) Rotation rate from −150 to 150°/s; (**f**) Rotation rate from −200 to 200°/s.

The RMSE results are calculated before and after denoising in Table 2 and Table 3. It is observed that the AMA-RWE-DFAKF has the minimum RMSE compared with other algorithms. Thus the effectiveness of this improved algorithm is verified in denoising FOG signal under both static and dynamic conditions.

Table 2. RMSE results of ±50°/s FOG signal with sampling 100 HZ.

Rotation (°/s)	Input	CKF	RWE-AKFG	AMA-RWE-DMAKF	AMA-RWE-DFAKF
+50	0.0384	0.0055	0.0040	0.0040	0.0040
+20	0.0462	0.0071	4.6724	0.0056	0.0038
+10	0.0573	0.0067	0.7951	0.0055	0.0036
+8	0.1145	0.0064	0.0103	0.0064	0.0032
+6	0.0515	0.0069	0.0071	0.0056	0.0037
+4	0.0403	0.0071	0.0095	0.0054	0.0038
+2	0.0358	0.0070	0.0097	0.0052	0.0037
0	0.0351	0.0073	0.0087	0.0054	0.0040
0	0.0349	0.0072	0.0039	0.0053	0.0039
−2	0.0366	0.0073	0.0102	0.0055	0.0040
−4	0.0441	0.0071	0.0102	0.0055	0.0046
−6	0.0657	0.0069	0.0102	0.0058	0.0037
−8	0.1552	0.0069	0.0117	0.0075	0.0040
−10	0.0778	0.0068	0.0071	0.0060	0.0037
−20	0.0402	0.0072	0.7679	0.0055	0.0039
−50	0.0369	0.0072	4.6649	0.0054	0.0039

Table 3. RMSE results of ±200°/s FOG dynamic signal with sampling 200 HZ.

Rotation(°/s)	Input	CKF	RWE-AKFG	AMA-RWE-DMAKF	AMA-RWE-DFAKF
0	0.0566	0.0087	0.0056	0.0056	0.0055
−2	0.0633	0.0165	0.0138	0.0082	0.0063
+2	0.0637	0.0163	0.0137	0.0081	0.0062
−4	0.1208	0.0233	0.0221	0.0087	0.0067
+4	0.1215	0.0234	0.0237	0.0150	0.0075
−6	0.1375	0.0205	0.0809	0.0186	0.0075
+6	0.1393	0.0208	0.1616	0.0187	0.0083
−8	0.1619	0.0210	0.2794	0.0176	0.0081
+8	0.1635	0.0204	0.4068	0.0170	0.0072
−10	0.1694	0.0215	0.5742	0.0175	0.0079
+10	0.1681	0.0197	0.7436	0.0163	0.0074
−20	0.1317	0.0239	1.8871	0.0213	0.0085
+20	0.1300	0.0226	3.3568	0.0201	0.0081
−50	0.1327	0.0174	8.8623	0.0156	0.0067
+50	0.1214	0.0161	15.5376	0.0153	0.0066
−80	0.1138	0.0118	23.1207	0.0107	0.0059
+80	0.1082	0.0117	31.3297	0.0109	0.0059
−100	0.1000	0.0114	37.0857	0.0098	0.0054
+100	0.0958	0.0111	43.3255	0.0096	0.0051
−150	0.0972	0.0100	59.1451	0.0096	0.0053
+150	0.0968	0.0102	76.2005	0.0097	0.0052
−180	0.1041	0.0098	86.9188	0.0086	0.0052
+180	0.1049	0.0099	97.8416	0.0087	0.0054
−200	0.1107	0.0096	105.2709	0.0085	0.0049
+200	0.1106	0.0109	112.7862	0.0088	0.0053
0	0.0563	0.0103	43.2174	0.0087	0.0055

Take the rotation rate from −50°/s to +50°/s as an example, the window size of RWE in AMA-RWE-DFAKF is discussed shown in Figure 12, because it can affect the convergence speed and filtering accuracy. The window of size is set to 5, 25, and 55, denoting small, middle, and long window. It can be seen that the smaller window size (N = 5) gives the faster convergence speed shown in Figure 12a, but has the poor filtering accuracy shown in Figure 12b. While, the longer window size (N = 55) give the higher filtering accuracy, but has the poor convergence speed. To get a relatively higher accuracy and faster convergence speed, the window size (N = 25) is chosen. The filter's convergence speed shown in Figure 13 is about 0.2 s for AMA-RWE-DFAKF, 0.4 s for AMA-RWE-DMAKF, 1.2 s for CKF, and 30 s for RWE-AKF. Moreover, the proposed filter's convergence speed is also faster compared with that of CKF, RWE-AKF, and AMA-RWE-DMAKF when the rotation rate is varying in other different values such as ±100°/s, ±150°/s and ±200°/s.

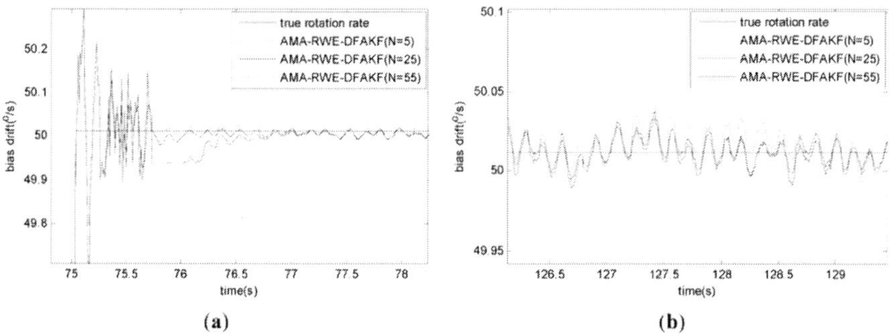

Figure 12. Comparison of denoising results for FOG signal with different parameters. (**a**) Comparison of convergence speeds; (**b**) Comparison of filtering accuracy.

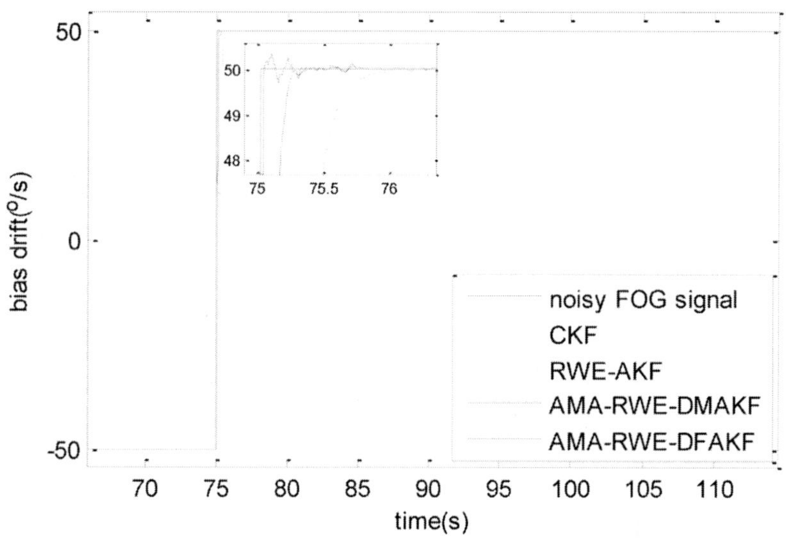

Figure 13. Comparison of denoising results for FOG dynamic signal.

6. CONCLUSIONS

In this paper, an AMA-RWE-DFAKF algorithm is proposed to denoise FOG static and dynamic signals. AMA is used to detect the discontinuities and RWE is used to estimate the double-factor in KF. The first adaptive parameter is Kalman gain updated by using RWE of the covariance matrix of innovation sequence. The second adaptive parameter is the covariance matrix of predicted state vector introduced to decrease the inertia at the discontinuities. In static condition, the performance of AMA-RWE-DFAKF is competitive with RWE-AKFG and AMA-RWE-DMAKF, but superior to CKF. Based on Allan variance analysis, the random errors like angle random walk and bias instability are reduced by 100 times. In dynamic condition, the minimum RMSE obtained by AMA-RWE-DFAKF show that it performs better than all considered algorithms. The effectiveness of this method is validated in denoising the single-axis FOG signal in both static and dynamic conditions. The future work is to verify this filtering in three-axis FOG and implement it in hardware.

ACKNOWLEDGMENTS

The project is supported by the National Natural Science Foundation of China (61340044, 11202010) and the Fundamental Research Funds for the Central Universities (YWF-10-01-B30). The authors acknowledge Beijing Aerospace Times Optical-Electronics Technology Co., Ltd for providing the FOG static and dynamic data.

AUTHOR CONTRIBUTIONS

This paper is completed by the listed authors. G.Y. proposed the idea of how to denoise the FOG random error in software. Y.L. completed the modeling, data processing, and manuscript writing. M. L. and S.S. participated in the experiments and paper modification.

REFERENCES

1. Bergh, R.A.; Lefevre, H.C.; Shaw, H.J. An overview of fiber-optic gyroscopes. *J. Lightw. Technol.* **1984**, *2*, 91–107.
2. Nayak, J. Fiber-optic gyroscopes: from design to production. *Appl. Opt.* **2011**, *50*, 152–161.
3. Kurbatov, A.M.; Kurbatov, R.A. Methods of improving the accuracy of fiber-optic gyros.*Gyroscopy Navig.* **2012**, *3*, 132–143.

4. IEEE standard speciation format guide and test procedure for single-Axis interferometric fiber optic gyros. *IEEE Std.* **1998**.

5. Song, N.F.; Yuan, R.; Jin, J. Autonomous estimation of Allan variance coefficients of onboard fiber optic gyro. *J. Instrum.* **2011**, *6*, 09005.

6. Hu, K.B.; Liu, Y.X. Adaptive noise cancellation method for fiber optic gyroscope. *Procedia Eng.* **2012**, *29*, 1338–1343.

7. Sabat, S.L.; Giribabu, N.; Nayak, J.; Krishnaprasad, K. Characterization of fiber optics gyro and noise compensation using discrete wavelet transform. In Proceedings of the 2nd International Conference on Emerging Trends in Engineering and Technology (ICETET), Nagpur, India, 16–18 December 2009; pp. 909–913.

8. Qian, H.M.; Ma, J.C. Research on fiber optic gyro signal de-noising based on wavelet packet soft-threshold. *J. Syst. Eng. Electron.* **2009**, *20*, 607–612.

9. Dang, S.W.; Tian, W.F.; Qian, F. EMD- and LWT-based stochastic noise eliminating method for fiber optic gyro. *Measurement* **2011**, *44*, 2190–2193.

10. Gan, Y.; Sui, L.F.; Wu, J.F.; Wang, B.; Zhang, Q.H.; Xiao, G.R. An EMD threshold denoising method for inertial sensors. *Measurement* **2014**, *49*, 34–41.

11. Yang, G.L.; Liu, Y.Y.; Wang, Y.Y.; Zhu, Z.L. EMD interval thresholding denoising based on similarity measure to select relevant modes. *Signal Process.* **2015**, *109*, 95–109.

12. Hammon, R.L. An application of random process theory to gyro drift analysis. *IRE Trans. Aeronaut. Navig. Electron.* **1960**, *7*, 84–91.

13. Wang, L.D.; Zhang, C.X. On-line modeling and filter of high-precise FOG signal. *Opt. Electron. Eng.* **2007**, *34*, 55–58.

14. Zhou, Y.; Li, Y.; Zhang, F. Modeling method of fiber optic gyro based on AR model. *Aerosp. Control Appl.* **2011**, *37*, 55–58.

15. Zhang, F. Modeling study on random error of fiber optic gyro. *Appli. Mech. Mater.* **2013**, *239*, 167–171.

16. Wu, X.M.; Duan, L.; Chen, W.H. A Kalman Filter approach based on random drift data of fiber optic gyro. In Proceedings of the 6th Conference on Industrial Electronics and Applications (ICIEA), Beijing, China, 21–23 June 2011; pp. 1933–1937.

17. Zhang, K.Z.; Tian, W.F.; Qian, F. A novel adaptive filter mechanism for improving the measurement accuracy of the fiber optic

gyroscope in maneuvering case. *Meas. Sci. Technol.* **2007**, *18*, 2777–2782.

18. Narasimhappa, M.; Sabat, S.L.; Peessapati, R.; Nayak, J. An improved adaptive Kalman Filter for denoising fiber optic gyro drift signal. In Proceedings of the 2013 Annual IEEE India Conference (INDICON), Mumbai, India, 13–15 December 2013; pp. 1–6.

19. Narasimhappa, M.; Sabat, S.L.; Rangababu, P.; Sabat, S.L. A modified Sage-Husa adaptive Kalman Filter for denoising fiber optic gyroscope signal. In Proceedings of the 2012 Annual IEEE India Conference (INDICON), Kochi, India, 7–9 December 2012; pp. 1266–1271.

20. Gao, S.S.; Gao, Y.; Zhong, Y.M.; Wei, W.H. Random weighting estimation method for dynamic navigation positioning. *Chin. J. Aeronaut.* **2011**, *24*, 318–323.

21. Gao, S.S.; Zhong, Y.M.; Wei, W.H.; Gu, C.F. Windowing-based random weighting fitting of systematic model errors for dynamic vehicle navigation. *Inform. Sci.* **2014**, *282*, 350–362.

22. Gao, S.S.; Hu, G.G.; Zhong, Y.M. Windowing and random weighting-based adaptive unscented kalman filter. *Int. J. Adapt. Control* **2015**, *29*, 201–223.

23. Narasimhappa, M.; Sabat, S.L.; Peessapati, R.; Nayak, J. An innovation based random weighting estimation mechanism for denoising fiber optic gyro drift signal. *Optik* **2014**, *125*, 1192–1198.

24. Karthik, K.P.; Rangababu, P.; Sabat, S.L.; Nayak, J. System on chip implementation of adaptive moving average based multiple-mode Kalman Filter for denoising fiber optic gyroscope signal. In Proceedings of the 2011 International Symposium on Electronic System Design (ISED), Kochi, Kerala, 19–21 December 2011; pp. 170–175.

25. Rangababu, P.; Sabat, S.L.; Karthik, K.; Nayak, J.; Giribabu, N. Efficient hybrid kalman filter for denoising fiber optic gyroscope signal. *Optik* **2013**, *124*, 4549–4556.

26. Rangababu, P.; Sabat, S.L.; Anumandla, K.K.; Kandyala, P.K.; Nayak, J. Design and implementation of realtime co-processor for denoising fiber optic gyroscope signal. *Digit. Signal Process.* **2013**, *23*, 1813–1825.

27. Kownacki, C. Optimization approach to adapt Kalman Filters for the real-time application of accelerometer and gyroscope signal' filtering. *Digit. Signal Process.* **2011**, *21*, 131–140.

28. Gao, W.W.; Wang, G.L.; Zhang, C.X.; Chen, J.H.; Gao, F.Q. An AMA-DWT-DMKF method for fiber optic gyroscope signal filtering. *Chin. J. Lasers* **2014**, *41*, 1–7.

29. Narasimhappa, M.; Sabat, S.L.; Nayak, J. An improved adaptive unscented Kalman Filter for denoising the FOG signal. In Proceedings of the 2014 Annual IEEE India Conference (INDICON), Pune, India, 11–13 December 2014; pp. 1–6.

30. Narasimhappa, M.; Sabat, S.L.; Nayak, J. An improved adaptive square root unscented Kalman Filter for denoising IFOG signal. In Proceedings of the 2014 IEEE International Symposium on Intelligent Signal Processing and Communication Systems (ISPACS), Kuching, Malaysia, 1–4 December 2014; pp. 159–164.

31. Narasimhappa, M.; Sabat, S.L.; Nayak, J. Adaptive sampling strong tracking scaled unscented kalman filter for denoising the fiber optic gyroscope drift signal. *IET Sci. Meas. Technol.* **2015**, *9*, 241–249.

Chapter 3

Integrated WiFi/PDR/Smartphone Using an Unscented Kalman Filter Algorithm for 3D Indoor Localization

Guoliang Chen [1], Xiaolin Meng [2,*], Yunjia Wang [1], Yanzhe Zhang [1], Peng Tian [1] and Huachao Yang [1]

[1]*School of Environment Science and Spatial Informatics, China University of Mining and Technology, 1 Daxue Road, 221116 Xuzhou, China*

[2]*Nottingham Geospatial Institute, University of Nottingham, Triumph Road, NG7 2TU Nottingham, UK*

ABSTRACT

Because of the high calculation cost and poor performance of a traditional planar map when dealing with complicated indoor geographic information, a WiFi fingerprint indoor positioning system cannot be widely employed on a smartphone platform. By making full use of the hardware sensors embedded in the smartphone, this study proposes an integrated approach to a three-dimensional (3D) indoor positioning system. First, an improved K-means clustering method is adopted to reduce the fingerprint database retrieval time and enhance positioning efficiency. Next, with the mobile phone's acceleration sensor, a new step counting method based on auto-correlation analysis is proposed to achieve cell phone inertial navigation positioning. Furthermore, the integration of WiFi positioning with Pedestrian Dead Reckoning (PDR) obtains higher positional accuracy with the help of the Unscented Kalman Filter algorithm. Finally, a hybrid 3D positioning system based on Unity 3D, which can carry out real-time positioning for targets in 3D scenes, is designed for the fluent operation of mobile terminals.

Keywords: Indoor localization; WiFi/PDR; Clustering; Auto-correlation analysis; Unscented Kalman Filter; Unity 3D

1. INTRODUCTION

Recently, a variety of indoor positioning technologies have emerged, such as ultrasound [1], infrared [2], Wireless Local Area Network, Bluetooth, Radio Frequency Identification [3], and Ultra Wideband. The RADAR [4], LANDMARC, and Place Lab techniques are the most representative within those approaches. However, these systems require additional hardware devices and complicated deployment. With the development of intelligent mobile terminals (such as smartphones and tablet computers), these devices can provide additional unprecedented functions. In addition, they can also provide a number of advanced technologies, such as WiFi, Bluetooth, and inertial sensors. A developer can make full use of these technologies and hardware to invent an indoor positioning system that is much easier to operate and less costly.

WiFi fingerprint indoor positioning systems focus on improving positioning accuracy and real-time performance [5,6,7]. There are two main methods to improve positioning accuracy, improving the positioning algorithm [8] or increasing the density of the fingerprint [9]. However, increasing the density of the fingerprint reduces the real-time performance because more fingerprint matching time is required whether based on direct interpolation positioning [10] or probability distribution positioning [11]. Especially when running on a mobile terminal, the time consumed by WiFi fingerprint positioning on limited hardware cannot meet real-time requirements. Therefore, improving the algorithm by lessening the amount of calculation is key to improving the effectiveness and positioning accuracy. Clustering the fingerprint database [12] is a good way to control the scale of the sample point search. The employment of a classical K-means clustering algorithm greatly assists the sub block processing of the WiFi fingerprint. With clustering, the time consumed by positioning is sharply decreased while the system precision is ensured, which effectively improves the real-time performance.

Pedestrian Dead Reckoning (PDR) positioning along with the use of the Inertial Measurement Unit that is built into a smartphone is widely used in navigation applications in indoor environments. Using PDR, a pedestrian's next position can be calculated when the starting position, heading information, and displacement are known. Many researchers use a step length model combined with direction information [13,14] to calculate the displacement of indoor pedestrians. However, the peak detection algorithm [15,16], which is commonly used for counting steps, cannot work well because of the large error of a smartphone. A new step counting method based on auto-correlation analysis of the sensor data is explored here. Compared with the traditional peak detection algorithm, it can clearly reduce the influence of step counting result errors caused by different mobile phone locations and motion postures of pedestrians.

PDR can provide motion information at a high update rate and achieve high precision over a short time duration. However, without external aids, the system suffers from local anomalies and cumulative error after positioning for a longer length of time. The positioning accuracy of WiFi fingerprinting depends on the density of the training samples, which means it does not have an accumulative error problem. Given the complementary characteristics of these two methods, combination of the two sources would bring better performance than a single source. Leppakoski [17] proposes Complementary Extented Kalman Filter (CEKF) for the fusion of PDR and WiFi positioning. Anshul [18] uses augmented particle filtering to simultaneously estimate location and user-specific walk characteristics, while the algorithm occupy large source and in open areas particle is less effective. Since the whole system is running on a resource limited smartphone, Chen zhenghua [19] formulates sensor fusion problem in a linear perspective and apply Kalman filter instead. However, the linearization of model is limited, the author attempts to integrate them with Unscented Kalman Filtering (UKF) algorithm. UKF is a nonlinear filtering algorithm proposed by Julier and based on the Unscented (U) transform. It can overcome the limitations of lower filter precision-caused truncation error and the necessity for a Jacobian matrix, and has been widely applied in nonlinear estimation [20].

With the development of the indoor positioning technology, there are a few of apps in initial stage. Xunlu is the first indoor navigation application for airports and markets in China which is based on WiFi positioning. However, it is easily influenced by the environment such as crowds which need multiple WiFi routers and adjust the relationship of WiFi signal intensity and location. Ibeacon is for IOS(iphone Operation System) based on BLE (Bluetooth Low Energy) location technology. Besides, IndoorAtlas developed an IndoorAtlas Application utilizing geomagnetic map, rather than WiFi or BLE. All these apps use single indoor positioning technology while no integration of different location methods. Most indoor positioning systems currently provide location services based on two-dimensional (2D) maps. The three-dimensional (3D) navigation and location system first appeared in the 2010 World Exposition in Shanghai [21]. Traditional 2D map is the abstraction of the indoor geographic information and cannot effectively express indoor geographic information because of the complexity of indoor environments, so it makes those users who are not professional difficult to identify his location. While 3D map express the actual three-dimensional space of indoor scene intuitively and realistic and show details of complex indoor environment, which could help people quickly figure out where he is and the direction he face. Moreover, smartphones cannot process the large amount of calculation needed for three-dimensional (3D) visualization. As a result, 3D visualization in mobile terminals has become an urgent problem to be solved. Mobile terminal 3D visualization relies on the Unity 3D platform, which enhances the expressiveness of indoor geographic information and user experience. The goal of this study is to design and develop a 3D indoor positioning system that locates indoor targets using a mobile terminal and monitors these indoor targets in a 3D scene.

The rest of this paper is organized as follows: Section 2 introduces WiFi fingerprint positioning and clustering of fingerprint databases. Section 3 presents an experimental analysis of PDR with a new step counting method. Section 4 presents the fusion of WiFi fingerprint positioning and PDR positioning using the UKF algorithm to improve accuracy. Section 5 shows how to achieve the 3D visualization in a mobile terminal using Unity 3D. Section 6 concludes and discusses the study's contributions.

2. WIFI FINGERPRINTING POSITIONING

A typical approach of WiFi fingerprint positioning consists of an offline phase and online phase. The offline phase is the first step to obtaining several access points (APs) that maintain a certain distance from each other in the positioning area. The data of each AP forms the sample point sets, denoted as $L = \{l_1, l_2, \ldots, l_m\}$. Each element in L consists of two parts: $Pi = (xi, yi)$ denotes the coordinates of each location and $Vi = [vi, 1, vi, 2, \ldots, vi, n]$ is the Received Signal Strength (RSS) vector received from the APs at the i-th reference point. The final database is generated by clustering using the K-means algorithm. In the online phase, the tracking target collects real-time RSS from APs and matches them with the fingerprints stored in the database to obtain the final position. The overall positioning process is shown in Figure 1.

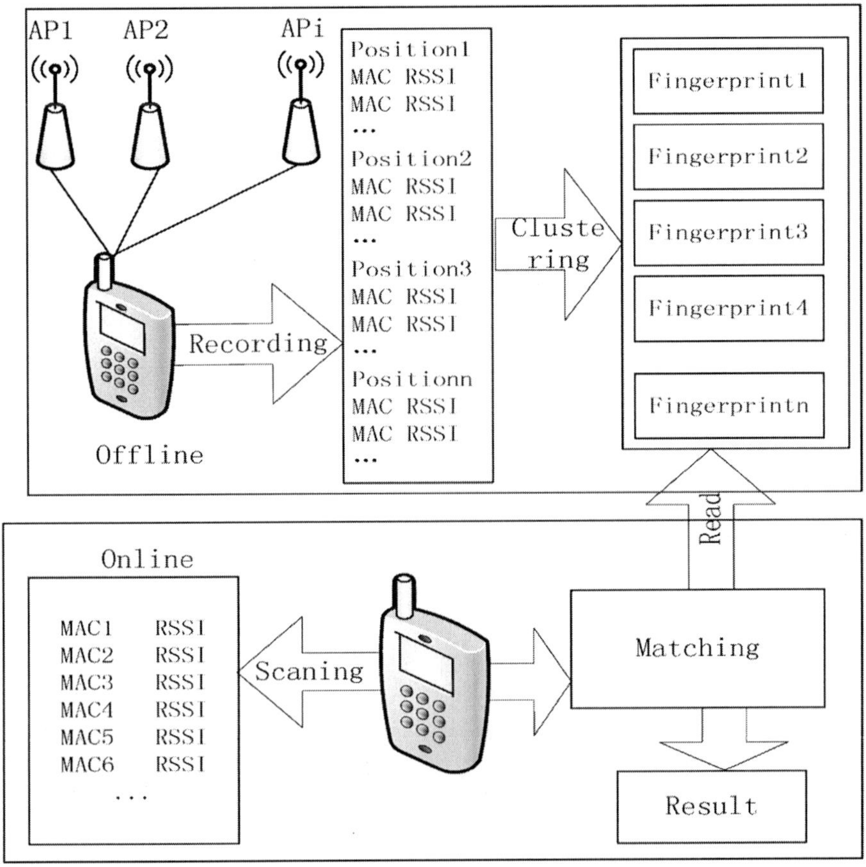

Figure 1. Positioning process.

2.1. WiFi Fingerprint Clustering

The primary clustering algorithms at present are the partitioning, hierarchical, density-based, grid-based, and model-based methods. The most typical method is the K-means clustering algorithm, which is a partitioning-type clustering method. Given the criterion for classification and the number of clusters K, this algorithm can divide a database of n data objects into K parts. The data within one cluster are extremely similar but are clearly different to those of other clusters.

WiFi fingerprint clustering divides sample points into several groups based on the signal distance between each two sample points. The distance is calculated as follows:

$$SD(l_i, l_j) = \|v_i, v_j\| = \sum_{k=1}^{n} \sqrt{(v_{i,k} - v_{j,k})^2} \tag{1}$$

The K-means algorithm clusters the fingerprint database as follows: First, k points are randomly picked as the initial center for each cluster. According to the distance from each center of the other clusters, the rest of the sample points are attached to the nearest cluster. The average value of each cluster is then recalculated. This recalculation repeats until the distance between each sample point and its clustering center are minimized. The calculation of E is as follows:

$$E = \sum_{i=1}^{k} \sum_{l \in C_i} \|l - c_i\| \tag{2}$$

where c_i is the average value of cluster C_i.

2.2. WiFi Fingerprint Positioning Algorithm

The traditional algorithm for WiFi fingerprint positioning is K Nearest Neighbor in Signal Space (K-NNSS), proposed by Bahl [4]. When positioning begins, real-time RSS measurements from the APs are collected to form vector $SV = [v_1, v_2, \ldots, v_n]$, $v_i (1 \leq i \leq n)$. The similarity $Length_i$ to each sample point in the fingerprint database is calculated as follows:

$$Length_i = \|SV, v_i\| = \sum_{k=1}^{n} \sqrt{(v_k - v_{i,k})^2} \tag{3}$$

All values of *Length* are ranked in ascending order and the *k* minimum *Length* coordinates are selected to form a vector $RL = \{RL_1, RL_2, \ldots, RL_k\}$, $(1 \leq i \leq k, RL_i = (x_i, y_i))$. The final positioning result is as follows:

$$RL = \sum_{j=1}^{k} w_j RL_j \tag{4}$$

where w_j is the weight of coordinate *j*.

The location area is too large for a sampling point to receive all AP signals, so APs which different sampling points in the fingerprint database can receive are different. The number of the same APs two sampling points can received signals from is larger, the two sampling points is closer. Therefore, the number of the same APs described above reflects spatial distance relationships of the two sampling points. Hence, taking the factor of spatial distance into the similarity calculation, the number of the same APs is utilized to improve similarity shown as below:

$$Length_i = (1 - \frac{NUM_s}{NUM}) \sum_{k=1}^{n} \sqrt{(v_k - v_{i,k})^2} \quad (5)$$

Here, *NUM* is the number of APs in the sampling area, and NUM_s is the number of these APs that can be sensed at the sampling point at the same time.

2.3. Experiment and Results

2.3.1. Experimental Setup

The corridor of the fourth floor corridor in the School of Environment Science and Spatial Informatics, China University of Mining and Technology was chosen as the test area. The building's general WiFi signal was used as the ubiquitous signal for positioning. There were 216 available APs in this 670 m^2 area.

This experiment consisted of a sampling phase, offline phase, and positioning phase. A Samsung Galaxy S III (Android 4.0) (Samsung Electronics, Seoul, Korea) was used as the device for sampling and positioning, and the sampling rate was 1 Hz. The RSS value received from each AP was an integer between −110 and 15. Each sampling point continuously offered 25 samples, and researchers recorded their average signal strength. In the offline phase, 390 available sampling points were collected (the average sample interval was 3.5 m) and stored in a database as fingerprints, which were processed using the K-means clustering algorithm. In the online positioning phase, 15 points were randomly chosen as test points and each were located 10 times to verify the algorithm's accuracy. The distribution of the sampling and test points are shown in Figure 2:

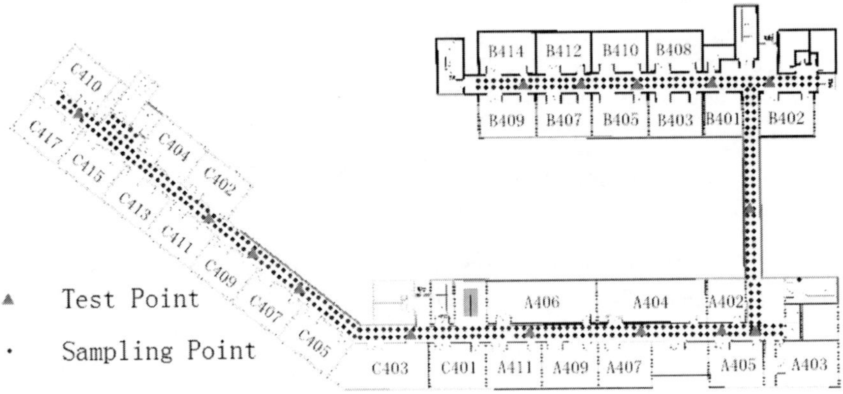

Figure 2. Distribution of sampling and test points.

2.3.2. Clustering Test

Three hundred and ninety sampling points were marked in sequence and configured as a data matrix (390 × 212) using Equation (3). The K-means algorithm is a supervised learning method and needs to be given the cluster number and initial center point. In this test, the cluster number was nine and the initial center was randomly picked. The clustering results are shown in Figure 3:

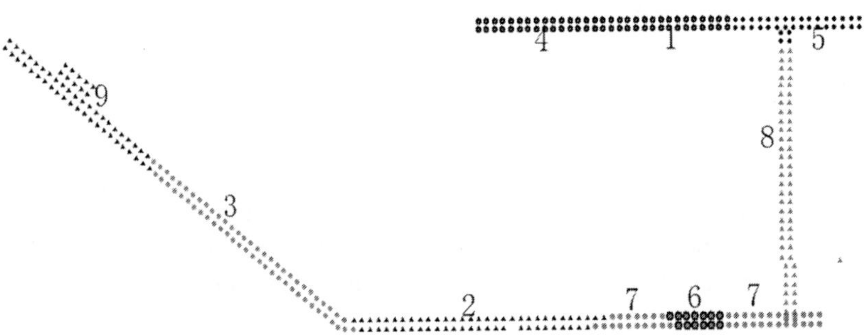

Figure 3. Clustering results.

The number of sampling points in each cluster is listed in Table 1.

Table 1. Number of sampling points in each cluster.

Class Number	1	2	3	4	5	6	7	8	9
Point Number	27	58	62	35	36	13	43	61	55

According to the results, the indoor environment clearly affects the clustering result, and the boundaries of different indoor environments are relevant to the boundaries of the clusters.

2.3.3. Positioning Test

The K-NNSS algorithm was used for positioning, where K was set to four. Considering that the number of repeated APs can reflect the spatial relationship, the improved similarity calculation in Equation (5) was selected to calculate *Length*. The Samsung Galaxy Ⅲ (Android 4.0) was also chosen for the positioning test. Fifteen randomly selected test points were continuously located 10 times. A comparison between clustered WiFi fingerprints and unclustered WiFi fingerprint was made with respect to three factors: average positioning time, average error, and maximum error.

Clustered WiFi fingerprints can reduce the positioning time by reducing the amount of data retrieved. As Figure 4 shows, the system's positioning time sharply decreases; the average reduction is 51%, the maximum reduction is 64%, and the minimum reduction is 36%.

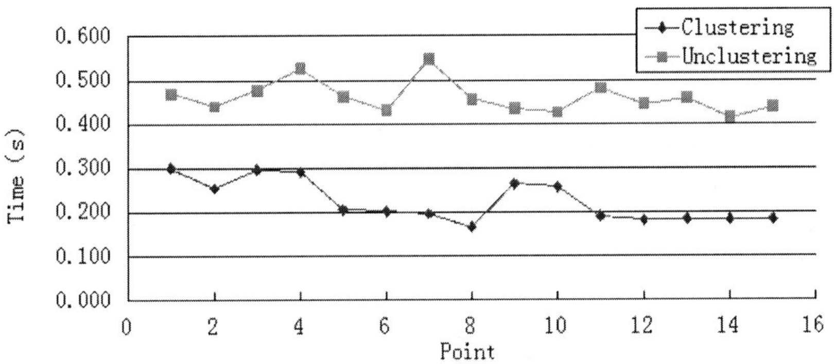

Figure 4. Single-point positioning time.

As Figure 5 and Figure 6 show, the system's average and maximum errors indicate that the single-point positioning results are more accurate after clustering. The average error was decreased from 1.92 to 1.12 m. The largest improvement was a reduction from 4.09 to 0.59 m. The average maximum error of single-point positioning decreased from 3.87 to 2.00 m, and the maximum reduction was 6.85 m. These results imply that the clustering algorithm can narrow the scope of the position search and reduce the error of the positioning result caused by instability of the WiFi signal.

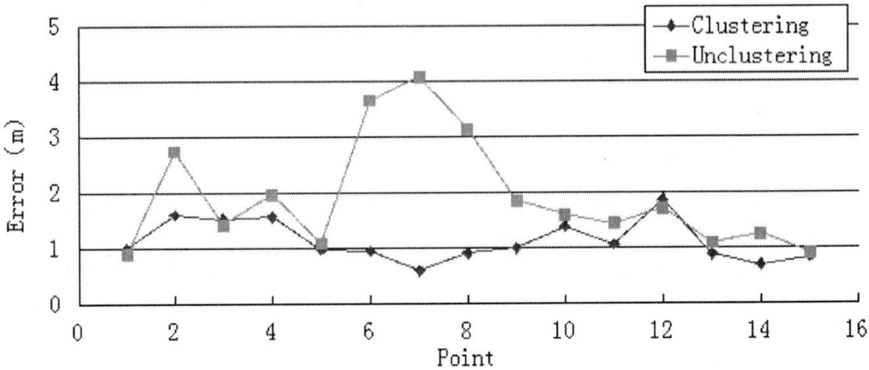

Figure 5. Single-point positioning average error.

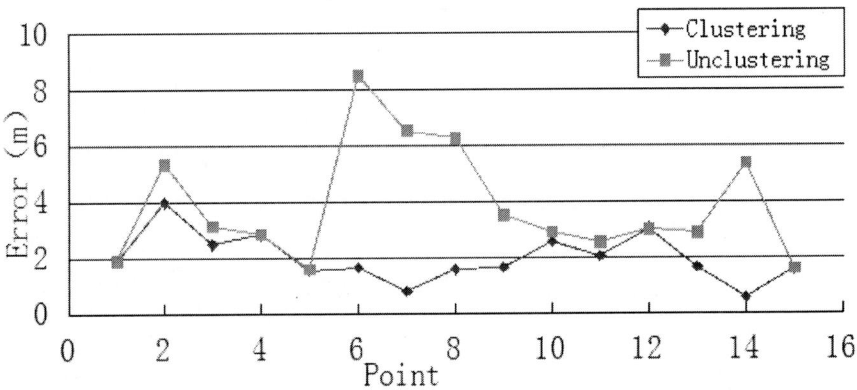

Figure 6. Single-point positioning maximum error.

3. INERTIAL POSITIONING

Typically, a pedestrians' movement over a very short sampling time is regarded as rectilinear motion In PDR, a pedestrian's next position can be deduced from the starting position, heading information, and displacement. The principle of PDR is shown in Figure 7:

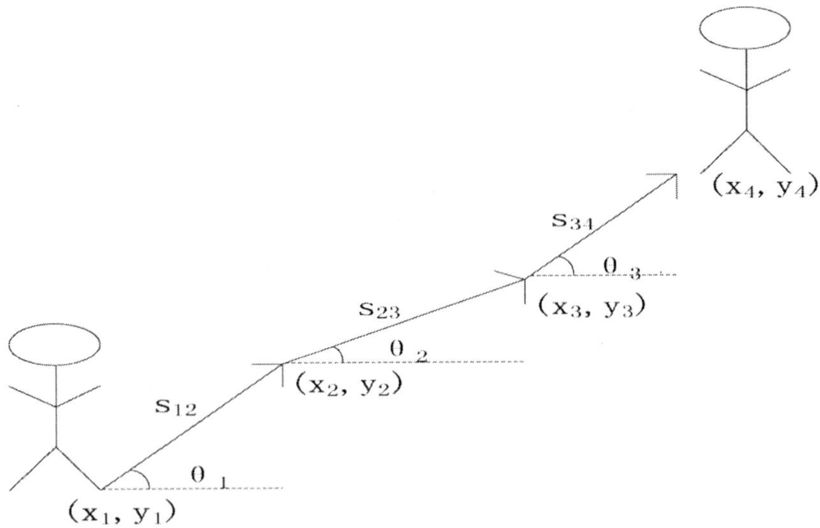

Figure 7. PDR principle.

If the initial position is assumed to be (x_i, y_i), the calculation of the next position (x_2, y_2) is:

$$y_2 = y_1 + S_{12} \cdot \cos\theta_1 \tag{6}$$

$$x_2 = x_1 + S_{12} \cdot \sin\theta_1 \tag{7}$$

Equations (6) and (7) can be easily adapted to obtain the next step's position. There are two key factors during the whole calculation: displacement *s* and heading direction ϑ. Of these, *s* can be estimated using a typical frequency-step model and ϑ can be obtained using an orientation sensor or gyroscope.

3.1. Counting Steps

The body's natural walking motion includes three components: forward, lateral, and vertical. These components with respect to a mobile phone's coordinate axes are shown in Figure 8.

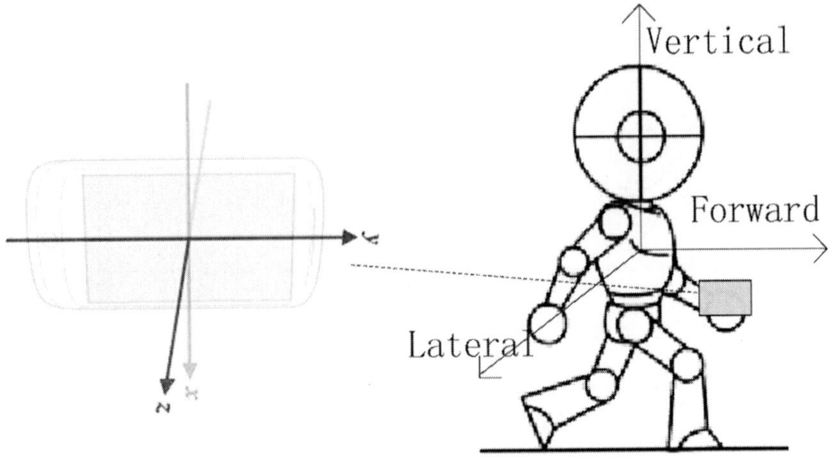

Figure 8. Schematic of coordinate axes.

The vertical acceleration in a step cycle changes regularly along with foot up-and-down movements. The acceleration changes along the three axes are shown in Figure 9. According to the clear periodicity of the *z*-axis acceleration, a step-counting algorithm based on the auto-correlation analysis of acceleration is proposed that can greatly reduce the error of step counting results caused by different orientations of the mobile phone and individual motion states.

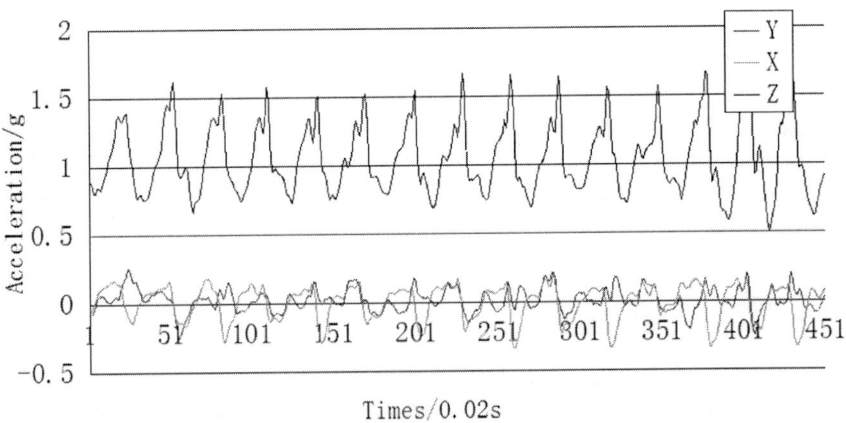

Figure 9. Three-axis acceleration of a mobile phone.

3.1.1. Auto-Correlation Analysis

The proposed auto-correlation analysis algorithm counts steps according to the similar acceleration of a pedestrian continuous motion. It divides the motion state into two cases: idle and walking. Stationary status includes the static state, standing up, sitting down, turning, gesturing, and other actions that do not change the body's position. The walking status occurs when the body changes its position while the mobile phone is being carried or used. There are two steps to calculating each step cycle: calculation of the standard deviation and calculation of the auto-correlation.

(1) Calculation of Standard Deviation

To reduce the influence of the phone's orientation, the overall acceleration is chosen to count steps. The overall acceleration described as follows:

$$a = \sqrt{{a_x}^2 + {a_y}^2 + {a_z}^2} \tag{8}$$

where a_x, a_y and a_z are the acceleration of the three axes. We identify the motion state using the standard deviation σ of a:

$$\sigma = \sqrt{\frac{\sum_{k=1}^{k=N} (a_k - u)^2}{N}} \tag{9}$$

where u is the mean of $\{a_1, a_2, \ldots, a_n\}$.

Figure 10 shows the Probability Density Function of the overall distribution of the acceleration standard deviation during a 1-s cycle, after recording walking and stationary statuses each for 5000 times. When the standard deviation is under 0.5, the probability that the pedestrian is stationary is greater than 99%. This statistic can be used as a heuristic threshold to judge the pedestrian's state, e.g., the pedestrian is considered to be stationary when the standard deviation is under 0.5. However, this value is not accurate enough to be used as a threshold for other states such as turning or some basic gestures. These motions have a larger standard deviation. To deal with this situation, an auto-correlation calculation is a better method to use.

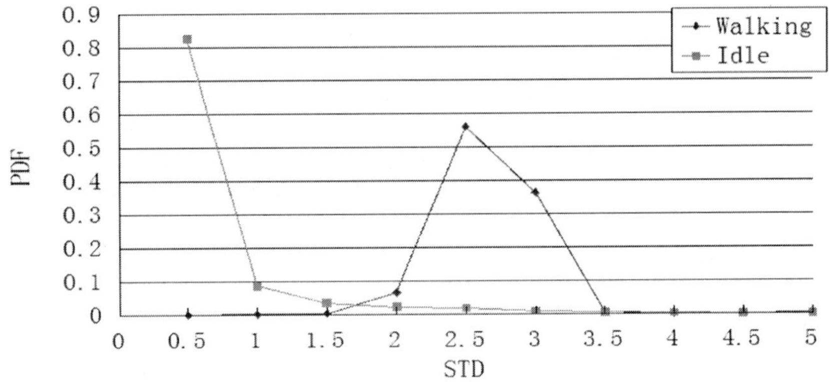

Figure 10. Distribution of the standard deviation of the overall acceleration in stationary and walking states.

(2) Calculation of Auto-correlation

According to Figure 11, the acceleration periodically changes because of the pedestrian's cyclical step movements. This characteristic can be applied to determine the auto-correlation between the current and previous strides, and hence identify the walking state.

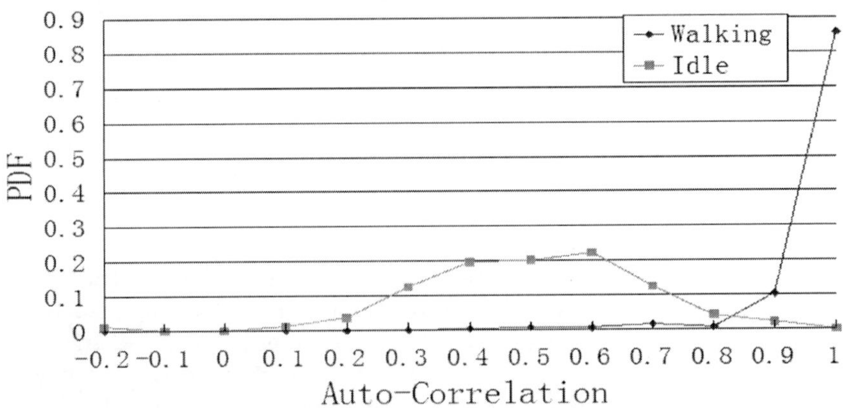

Figure 11. Distribution of auto-correlation during stationary and walking states.

The mobile phone acceleration sensor continuously records data when the pedestrian is walking. The overall auto-correlation of the acceleration is determined as follows:

$$x(m,t)=\frac{\sum_{k=0}^{k=t-1}[(a(m+k)-u(m,t))\times(a(m+k+t)-u(m+t,t))]}{t\times\sigma(m,t)\times\sigma(m+t,t)} \quad (10)$$

where $u(m, t)$ and $\sigma(m, t)$ are the mean and standard deviation of the sequence samples $\{a(k + 1), \ldots, a(k + t + 1)\}$. When the sampling period is close to the stride period, $x(m, t)$ is approximately one. However, different individuals or one individual at different times can have different stride frequencies, so t is a varied. The frame algorithm is designed to obtain t dynamically. The algorithm selects t from the range t_{min} – t_{max}, then calculates $\rho(m, t)$ as follows:

$$\rho(m,t)=\max_{t=t_{min}}^{t=t_{max}}(x(m,t)) \quad (11)$$

where t is the period of the acceleration pattern that $\rho(m, t)$ is maximized.

The stride frequency of normal walking ranges from 1 to 2.5 Hz, so t ranges from 0.4 to 1 s.Figure 11 presents the Probability Density Function of $\rho(m, t)$ for both stationary and walking state, each sampled 1000 times. Based on Figure 11, when $\rho(m, t)$ is 0.7 or higher, the probability of a walking state is greater than 90%.

3.1.2. Experiment and Results

According to the two thresholds for judging the motion state, auto-correlation algorithm is carried out. Setting t to 0.4–1.0 s and calculating the standard deviation of the overall sequence, the motion state was determined as follows:

(1) If $\sigma < 0.5$ then the state is stationary.

(2) If $\rho(m, t) > 0.7$ then the state is walking.

(3) Otherwise, the current state is unchanged.

Table 2 lists the test results for the auto-correlation algorithm and the peak detection algorithm.

Table 2. Performance of the auto-correlation algorithm.

Position of Mobile Phone	Motion State of Pedestrian	Peak Detection Algorithm Steps	Auto-Correlation Algorithm Steps	True steps
In Hand	Changing frequency of walking	224	196	200
In Jacket pocket	Constant speed walking	233	202	200
In Hand	phone using when walking	204	198	200
In Hand	Constant speed walking	199	197	200
In Hand	Idle	0	0	0
Change positions	Idle	6	6	0

According to the chart, the auto-correlation algorithm greatly reduces the influence of different mobile phone positions and various pedestrian motions compared to the peak detection algorithm. It can hence be applied to the step counting task in all situations.

3.2. Step Length Estimation

A frequency model is chosen as the generic step model calculated as follows:

$$L_g = a \times f + b \tag{12}$$

where *f* is walking frequency, and *a* and *b* are coefficients. The step length is difficult to set because of the different height and weight of individuals. The coefficient values are adopted from Li, who obtained parameters from 4000 steps of 23 different people [22].

3.3. Heading Direction Estimation

Heading direction can be determined by compass or gyroscope. Magnetic direction can be directly obtained from a direction sensor (compass), which is easily disturbed by circumstances. The magnetic heading angle is calculated as:

$$h_{mag} = \arctan(\frac{m_y}{m_x}) \tag{13}$$

where m_x and m_y are the Earth's magnetic field components along *x* and *y* axes of the local frame.

The strapdown heading direction comes from the gyroscope angular velocity and initial direction. It is stable when faced with external disturbance, but accumulates more error. The strapdown heading direction can be determined by integrating the *z*-axis angular velocity of the gyroscope:

$$h_k = h_{k-1} + w_k dt \tag{14}$$

where h_k is current strapdown heading direction and w_k is the angular velocity at step *k*. To reduce the circumstance disturbances as well as accumulative error, this study adopts a combination of both approaches to determine the heading direction, calculated as follows:

$$a = (1 - W)h_k + Wh_{mag} \tag{15}$$

where *W* is the weight of the magnetic data. The value of *w* is taken from [23].

However, obvious error could exist in the data from the mobile phone sensors, or the phone might not be oriented along the motion direction of the user. Given this situation, this study uses map information to constrain the heading direction.

4. UKF

PDR has a high precision over a short time, but gradually accumulates error. In contrast, WiFi fingerprint positioning has no accumulative error. The combination of WiFi fingerprint and PDR positioning leads to a more stable indoor positioning over the long term.

Error in PDR is mainly caused by the random error of various sensors (accelerometer, gyroscope, and compass). Hence, the system's positioning accuracy and reliability can be improved effectively using an optimal estimation method. The Extended Kalman Filter (EKF) is a simple and popular formulation for nonlinear estimation [24,25,26]. However, it causes model error in the linearization and reduces estimation accuracy. In addition, a high-dimensional complicated model called the Jacobi Matrix must be calculated for the EKF. In the 1990s, Julier proposed the UKF method, which uses the U transform in the filtering process to improve accuracy.

4.1. UKF Model

Assume a system dynamic model and measurement model as follows:

$$X_k = f(X_{k-1}, u_{k-1}) + W_{k-1}$$
$$Z_k = h(X_k) + V_k \tag{16}$$

where f and h are the nonlinear vector functions, W_k and V_k are the process noise and measurement noise, respectively, which are both uncorrelated zero mean white Gaussian noise, and their covariances are Q_k and R_k, respectively. Finally, u_{k-1} is the control input of the model.

A UKF is a Kalman Filter based on the U transform. Suppose an N-dimensional random variable$X \sim N(X^{-}, P_X)$. Further, Z is the statistic characteristic (Z^{-}, P_Z) transformed from X according to a nonlinear function $f(\cdot)$. The U transform designs a series of points $\xi_i(i = 1, 2, ..., L)$ named Sigma points using (X^{-}, P_X), and calculates the result $\chi_i(i = 1, 2, ..., L)$ of the Sigma points using $f(\cdot)$ to obtain the result (Z^{-}, P_Z) based on χ_i. Generally, the number of Sigma points is 2n + 1. The U transform can be described as follows:

(1) Calculation of Sigma points

$$\begin{cases} \xi_0 = \overline{X} \\ \xi_i = \overline{X} - (\sqrt{(n+\lambda)P_X})_i, (i = n+1, n+2, \cdots, 2n) \\ \xi_i = \overline{X} - (\sqrt{(n+\lambda)P_X})_i, (i = n+1, n+2, \cdots, 2n) \end{cases} \tag{17}$$

$$\begin{cases} \omega_0^{\mathrm{m}} = \dfrac{\lambda}{n+\lambda} \\ \omega_0^{\mathrm{c}} = \dfrac{\lambda}{n+\lambda} + (1 - \alpha^2 + \beta) \\ \omega_i^{\mathrm{m}} = \omega_i^{\mathrm{c}} = \dfrac{1}{2(n+\lambda)}, i = 1, 2, \cdots, 2n \end{cases} \tag{18}$$

$$\lambda = \alpha^2 (n+k) - n \tag{19}$$

where α is usually a small positive number such as 0.001, $k = 0$, β is usually used to describe the information distribution of X (the optimal value of β is two in a Gaussian noise environment). Further, $(\sqrt{(n+\lambda)P_X})_i$ represents column *i* of the square root matrix, and ωmi and ωci are the weight coefficients for calculating first- and second-order statistical characteristics.

(2) Calculation of χ_i, P_{ZZ} and P_{XZ}

$$\chi_i = f(\xi_i), i = 0,1,\cdots,2n \tag{20}$$

$$\begin{cases} \overline{Z} = \sum_{i=0}^{2n} \omega_i^m \chi_i \\ P_{ZZ} = \sum_{i=0}^{2n} \omega_i^c (\chi_i - \overline{Z})(\chi_i - \overline{Z})^T \\ P_{XZ} = \sum_{i=0}^{2n} \omega_i^c (\xi_i - \overline{Z})(\chi_i - \overline{Z})^T \end{cases} \tag{21}$$

A flow chart of the UKF is presented in Figure 12.

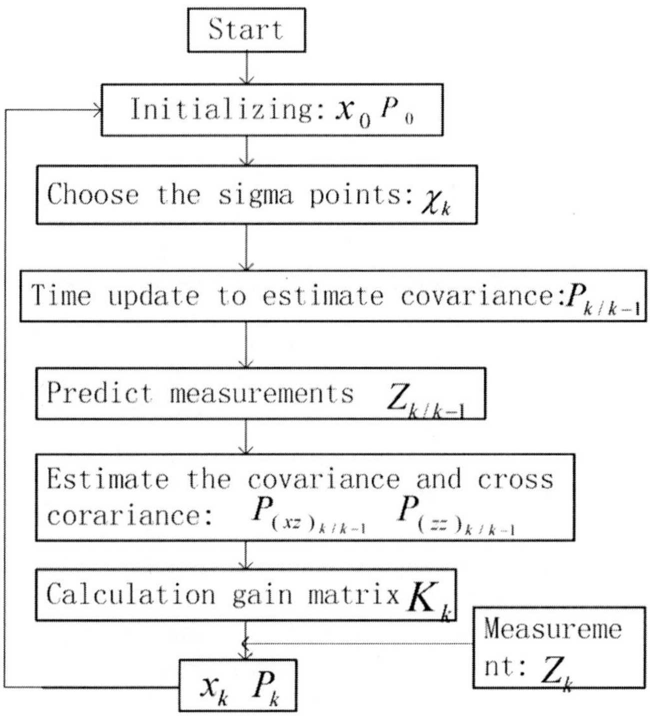

Figure 12. UKF process.

4.2. Implementation of UKF Fusion

UKF is used to fuse the results of WiFi fingerprint positioning and PDR positioning. The UKF fusion is achieved by directly using step domain instead of time domain, which could eliminate the process of converting the PDR location to the distance meter before fusion with WiFi positioning information, and reduce the complexity of the algorithm. According to the pedestrian state, a system model is built with the nonlinear formula as follows:

$$X_k = \begin{pmatrix} x_k \\ y_k \\ \theta_k \end{pmatrix} = \begin{pmatrix} x_{k-1} + \widetilde{s} \cdot \cos\theta_{k-1} \\ y_{k-1} + \widetilde{s} \cdot \sin\theta_{k-1} \\ \theta_{k-1} + \widetilde{\theta} \end{pmatrix} + W_{k-1} \quad (22)$$

where x_k and y_k are the position after k steps, ϑ_k is the heading direction after k steps, W_{k-1} is three-dimensional system process noise, s˜ is the step length of step k, and θ˜ is the variation of the heading direction in step k. The measurement model is created as follows:

$$Z_k = \begin{pmatrix} x_k \\ y_k \\ s_k \\ \Delta\theta_k \\ \theta_k \end{pmatrix} = \begin{pmatrix} x_k \\ y_k \\ \sqrt{(x_k - x_{k-1})^2 + (y_k - y_{k-1})^2} \\ \theta_k - \theta_{k-1} \\ \theta_k \end{pmatrix} + V_k \quad (23)$$

where x_k and y_k are the position of pedestrian from WiFi fingerprint positioning, s_k is length of step k from PDR positioning, $\Delta\vartheta_k$ is the variation of the heading direction of step k from the gyroscope, ϑ_k is the heading direction of step k from the compass, and V_k is five-dimensional system measurement noise.

4.3. Experiment and Results

The location experiment was carried out in zones A and B on the fourth floor of the School of Environment Science and Spatial Informatics, China University of Mining and Technology. The initial location of the pedestrian subject was known. The tester moved from the starting point shown in Figure 13 to the end after two right-angle corners, for a total of 133.5 m. The initial positioning performance coincided with the real path, but declined over time. According to Figure 13, the accumulative error reached 3.5 m at corner B, and reached 6.8 m at corner C. The WiFi fingerprint positioning has a large error and, in some cases, its position result concentrated on a small area. As the WiFi fingerprint positioning result was used for the filter initial value, the initial position error was obvious, about 2.3 m in corner A. As time went by, the fusion result lessened the error, which was just under 1 m in corner B and 1.7 m in corner C. Compared with the 6.8-m error of PDR positioning only, the error was greatly improved. When the tester reached end D after 133.5 m, the PDR positioning error was 6.0 m. In contrast, the fusion positioning error was just 1.5 m. From this experiment, it can be concluded that the UKF filter algorithm could fuse WiFi fingerprint positioning and PDR positioning well, and the correction of WiFi fingerprint positioning fixes the accumulative error of the PDR positioning.

In order to test the effectiveness of the UKF fusion algorithm, the author made a contrast test from the opposite side in the original test site. The tester moved from the starting point A after two right-angle corners B and C, until to the end point D, for a total of 138.3 m, 219 steps. As shown in Figure 14, The UKF fusion method improved location results effectively compared to PDR positioning and WiFi positioning alone.

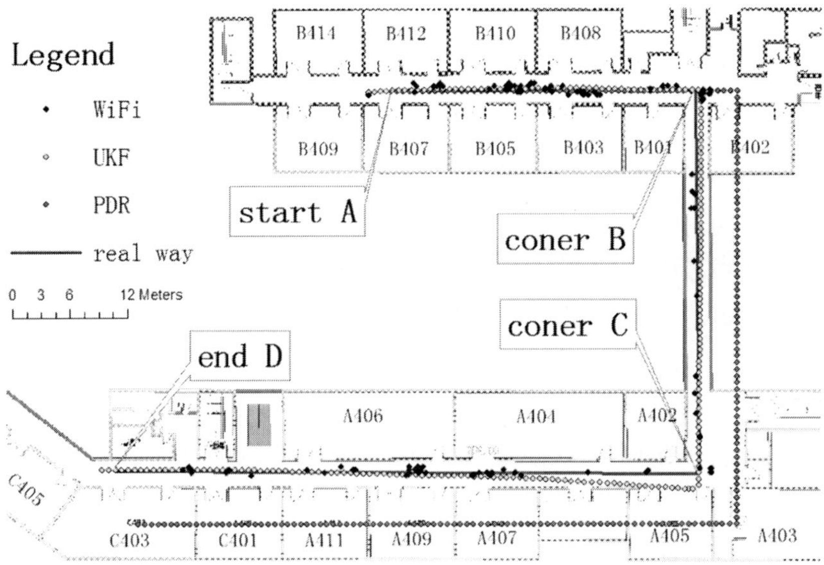

Figure 13. UKF positioning test.

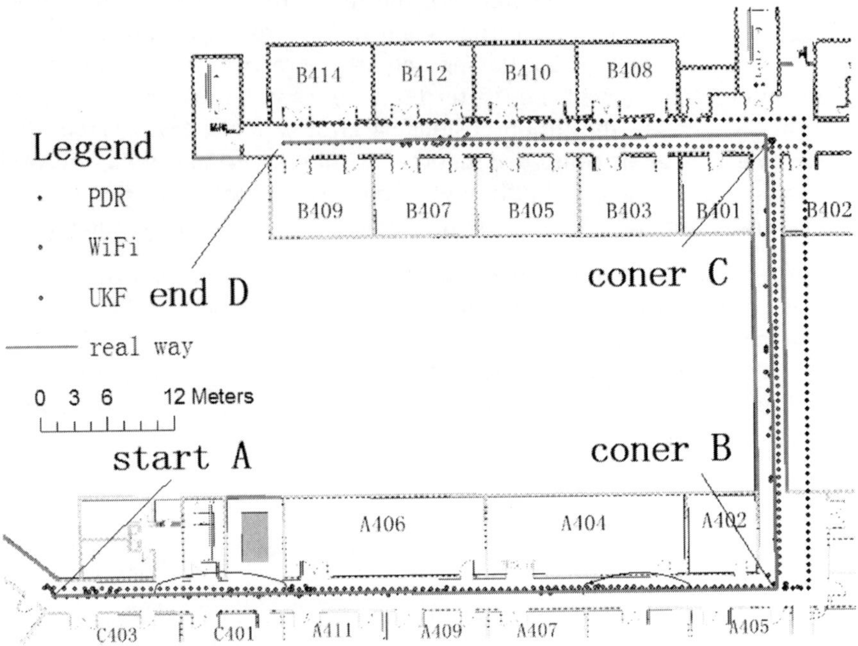

Figure 14. UKF positioning comparison test.

5. 3D INDOOR POSITIONING SYSTEM DESIGN AND IMPLEMENTATION

5.1. System Design

The proposed 3D indoor positioning system is divided into three modules: positioning, display, and network. The positioning module contains the localization algorithm and obtains a person's position. The display module includes the mobile and monitor modules. The mobile module displays the positioning result on the 3D scene on the mobile terminal and the monitor module displays all users' positioning results in the 3D scene at the monitoring terminal. The network module is also further divided into mobile and monitor modules. The mobile module sends the user's position information to the monitor module and the monitor module receives all online users' positioning information and saves them into a database. The structure of the system is shown in Figure 15.

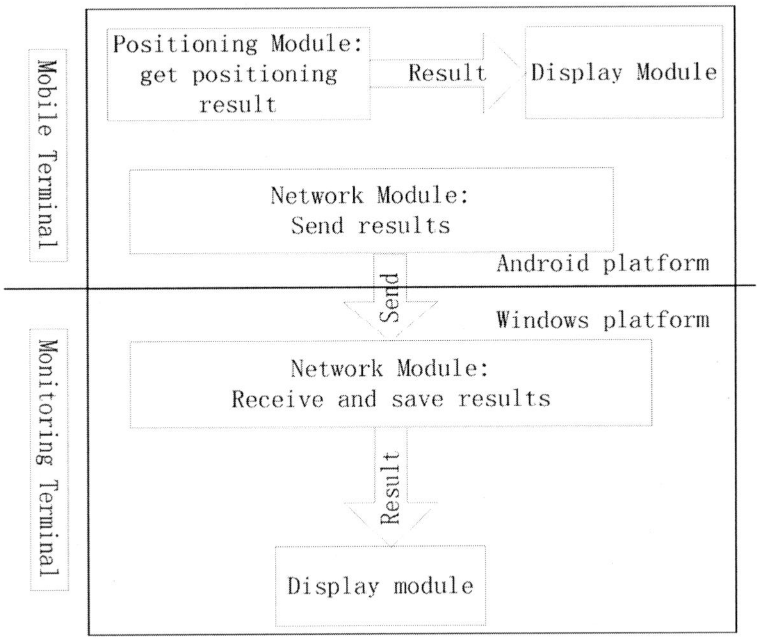

Figure 15. Structure of the 3D indoor positioning system.

5.2. System Implementation and Test

The implementation of 3D Indoor Positioning System is based on Unity 3D platform which can generate the real-time 3D scene in the mobile terminal. The Samsung Galaxy III (Android 4), equipped with a WiFi module and 1433 MHz CPU, was used as the mobile terminal test platform. The monitor terminal platform was a Lenovo G480 notebook computer (Lenovo, Beijing, China) whose operating system was Windows 7 Home Basic 64 bit, and was equipped with an Intel Core i5 2.5 GHz CPU. Our test field was in the School of Environment Science and Spatial Informatics, as mentioned above, and the real scene of the fourth floor corridor test area and the 3D scene in the positioning system are shown respectively in Figure 16 and Figure 17.

In the mobile positioning terminal, users cannot only acquire their location within the whole building as a bird's eye view, but also switch to the current position indicated by the red point in the 3D scene, which changes to follow a user's path. Meanwhile, users can acquire each floor of the whole building and send position information to the monitoring terminal. The monitoring terminal is responsible for showing all online users' positions in the whole building and monitoring each floor. While tracking specified targets, it can switch 3D scenes to the appropriate monitoring floor and show the related information.

Figure 16. Indoor positioning test field.

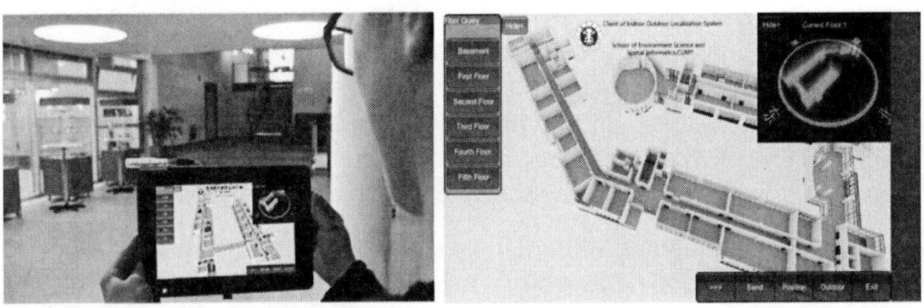

Figure 17. Interface of 3D-indoor Positioning System.

6. CONCLUSIONS

In this work, we proposed an integration of WiFi fingerprinting positioning and PDR that uses a UKF algorithm to improve the positioning accuracy and reliability. For WiFi fingerprinting localization, the improved K-means clustering algorithm was proposed to reduce the resource cost of the location algorithm and improve the system's real-time performance without reducing the positioning accuracy. The time consumed was reduced on average by 51%, with the largest decline reaching 64% and the smallest value 36%. In the PDR approach, an auto-correlation analysis algorithm, which can greatly reduce the influence of mobile phone's position and pedestrian's motions on the result of step counting, was applied to count steps. To display the geographic information vividly, a 3D indoor positioning system based on the Unity 3D platform was devised. It can express complicated indoor space geographic information and enrich the user experience.

ACKNOWLEDGMENTS

This research project was jointly supported by the National Natural Science Foundation of China (No. 41371423), Ministry of Science and Technology of China (No. 2013AA12A201), Key Laboratory for Aerial Remote Sensing Technology of National Administration of Surveying, Mapping, and Geo-information(NASG) (No. 2014A01), and a Project Funded by the Priority Academic Program Development of Jiangsu Higher Education Institutions.

AUTHOR CONTRIBUTIONS

Guoliang Chen and Xiaolin Meng conceived and designed the experiments; Yanzhe Zhang performed the experiments; Peng Tian and Huachao Yang analyzed the data; Yunjia Wang contributed reagents/materials/analysis tools; G.C. wrote the paper.

CONFLICTS OF INTEREST

The authors declare no conflicts of interest.

REFERENCES

1. Gao, X.C.; Jiang, T.; Cao, L.F. Design of Indoor Location System Based on FRID. *J. Guangxi Acad. Sci.* **2013**, *29*, 31–33.

2. Wang, Y.; Lin, J.G. Indoor Location Technique Based on RFID. *China Instrum.* **2011**, *02*, 54–57.

3. Yang, Y.; Xiao, J.H.; Liu, Z. Indoor Three-Dimensional Positioning System Based on Ultrasonic. *J. Jilin Univ.* **2012**, *30*, 267–271.

4. Bahl, P.; Padmanabhan, V. RADAR: An in-building RF-based user location and tracking system. In Proceedings of the Nineteenth Annual Joint Conference of the IEEE Computer and Communications Societies, Tel Aviv, Israe, 26–30 March 2000; pp. 775–784.

5. Kuo, S.P.; Tseng, Y.C. Discriminant minimization search for large-scale RF-based localization systems. *IEEE Trans. Mob. Comput.* **2011**, *10*, 291–304.

6. Du, Y.; Yang, D.; Xiu, C. A Novel Method for Constructing a WIFI Positioning System with Efficient Manpower. *Sensors* **2015**, *15*, 8358–8381.

7. Tang, J.; Chen, Y.; Chen, L.; Liu, J.; Hyyppä, J.; Kukko, A.; Kaartinen, H.; Hyyppä, H.; Chen, R. Fast Fingerprint Database Maintenance for Indoor Positioning Based on UGV SLAM.*Sensors* **2015**, *15*, 5311–5330.

8. Li, K.J.; Bigham, J.; Bodanese, E.L.; Tokarchuk, L. Location estimation in large indoor multi-floor buildings using hybrid networks. In Proceedings of the IEEE Wireless Communications and Networking Conference (WCNC), Shanghai, China, 7–10 April 2013; pp. 2137–2142.

9. Zdruba, G.V.; Huber, M.; Kamanger, F.A.; Chlarntac, I. Monte Carlo sampling based in-home location tracking with minimal RF infrastructure requirements. In Proceedings of the IEEE Global Telecommunications Conference (GLOBECOM'04), Dallas, TX, USA, 29 November–3 December 2004; pp. 3624–3629.

10. Chai, X.Y.; Yang, Q. Reducing the calibration effort for probabilistic indoor location estimation. *IEEE Trans. Mob. Comput.* **2007**, *6*, 649–662.

11. Wymeersch, H.; Jaime, L.; Win, M.Z. Cooperative localization in wireless networks. In Proceedings of the IEEE, Guilin, China, 18–21 May 2009; pp. 427–450.

12. Kuo, S.P.; Wu, B.J.; Peng, W.C. Cluster-enhanced techniques for pattern-matching localization systems. In Proceedings of the IEEE International Conference on Mobile *Ad hoc*and Sensor Systems, Pisa, Italy, 8–11 October 2007; pp. 1–9.

13. Shih, W.Y.; Chen, L.Y.; Lan, K.C. Estimating walking distance with a smart phone [C]. In Proceedings of the 5th International Symposium on Parallel Architectures, Algorithms and Programming (PAAP), Taipei, Taiwan, 17–20 December 2012; pp. 166–171.

14. Subramanian, S.P.; Sommer, J.; Zeh, F.P. PBIL-PDR for scalable Bluetooth Indoor Localization. In Proceedings of the 3rd International Conference on Next Generation Mobile Applications, Services and Technologies, Cardiff, Wales, 23–25 September 2009; pp. 170–175.

15. Lan, K.C.; Shih, W.Y. Using Smart-Phones and Floor Plans for Indoor Location Tracking.*IEEE Trans. Hum. Mach. Syst.* **2014**, *44*, 211–221.

16. Liu, J.B.; Chen, R.Z.; Pei, L.; Guinness, R.; Kuusniemi, H. A Hybrid Smartphone Indoor Positioning Solution for Mobile LBS. *Sensors* **2012**, *12*, 17208–17233.

17. Leppäkoski, H.; Collin, J.; Takala, J. Pedestrian navigation based on inertial sensors, indoor map, and WLAN signals. *J. Signal Process. Syst.* **2013**, *71*, 287–296.

18. Rai, A.; Chintalapudi, K.K.; Padmanabhan, V.N.; Sen, R. Zee: Zero-effort crowdsourcing for indoor localization. In Proceedings of the 18th annual international conference on Mobile computing and networking, Istanbul, Turkey, August 22–26 2012; pp. 293–304.

19. Chen, Z.; Zou, H.; Jiang, H.; Zhu, Q.; Soh, Y.C.; Xie, L. Fusion of WiFi, Smartphone Sensors and Landmarks Using the Kalman Filter for Indoor Localization. *Sensors* **2015**, *15*, 715–732.

20. Julier, S.J. A skewed approach to filtering. In Proceedings of the SPIE 3373, Signal and Data Processing of Small Targets 1998, 271, Orlando, FL, USA, 3 September 1998; pp. 271–282.

21. Chen, R.Z.; Kuusnieme, H.D.; Hyyppa, J.H.; Zhang, J.; Takala, J.; Kuittinen, R.; Wang, J. Going 3D Personal Nav and LBS. *GPS World* **2010**, *21*, 14–18.

22. Li, F.; Zhao, C.S.; Ding, G.Z.; Gong, J.; Liu, C.; Zhao, F. A Reliable and Accurate Indoor Location Method Using Phone Inertial Sensors. In Proceedings of the 2012 ACM Conference on Ubiquitous Computing, New York, NY, USA, 5–8 September 2012; pp. 421–430.

23. Kling, L.; Wark, T. A wireless sensor network for real-time indoor localisation and motion monitoring. In Proceedings of the International Conference on Information Processing in Sensor Networks, 2008, St louis, MO, USA, April 22–24 2008; pp. 39–50.

24. Fang, J.C.; Shen, G.X.; Wan, D.J. An Adaptive Federated Kalman Filter and its Application at GPS/DR Integrated Navigation System in Land Vehicle. *J. Chin. Inert. Technol.* **1998**, *6*, 1–6.

25. Wu, Q.P.; Wan, D.J.; Wang, Q. A New Kind of Filter Algorithm of Integrated Navigation System for Vehicle. *J. Chin. Inert. Technol.* **1999**, *7*, 22–24.

26. Fang, J.C.; Shen, G.X.; Wan, D.J. Adaptive Extended Kalman Filter Model for Integrated Land Vehicle Navigation System. *J. Beijing Univ. Aeronaut. Astronaut.* **1999**, *25*, 235–239.

Chapter 4

Kalman Filters versus Neural Networks in Battery State-of-Charge Estimation: A Comparative Study

Ala A. Hussein

Department of Electrical Engineering, UAE University, Al Ain, UAE

ABSTRACT

Battery management systems (BMS) must estimate the state-of-charge (SOC) of the battery accurately to prolong its lifetime and ensure a reliable operation. Since batteries have a wide range of applications, the SOC estimation requirements and methods vary from an application to another. This paper compares two SOC estimation methods, namely extended Kalman filters (EKF) and artificial neural networks (ANN). EKF is a nonlinear optimal estimator that is used to estimate the inner state of a nonlinear dynamic system using a state-space model. On the other hand, ANN is a mathematical model that consists of interconnected artificial neurons inspired by biological neural networks and is used to predict the output of a dynamic system based on some historical data of that system. A pulse-discharge test was performed on a commercial lithium-ion (Li-ion) battery cell in order to collect data to evaluate those methods. Results are presented and compared.

Keywords: Artificial neural network (ANN), battery, extended Kalman filter (EKF), state-of-charge (SOC)

1. INTRODUCTION

Cost, cycle-life, reliability, and safety are major concerns in all batteries. Due to their high cost in general, batteries must be carefully charged and discharged in order to prolong their cycle-life and thus minimize their maintenance and replacement costs. The battery must also be operated within its design limits to ensure a reliable and safe operation. That is, overcharging batteries produces excessive heat that could damage the battery, shorten its lifetime, and further, result in a fire hazard, where over-discharging batteries could result in an irreversible damage or reduced service-time [1] [2] . Thus, the main function of a battery management system (BMS) is to ensure that the battery operates within its design limits. One important parameter that must be continuously and accurately estimated, under any condition, to prevent the battery from exceeding its design limits is the SOC. Similar to the concept of a fuel gauge, an SOC estimator must provide an accurate value of the amount of energy in the battery as a percentage of its total capacity [3].

Unlike a fuel gauge, the SOC is an immeasurable quantity that is rather estimated. Nonetheless, there are several methods that can be used to estimate the SOC ranging from simple to sophisticated methods. Those methods can be classified into two types: direct and indirect. Some examples of direct methods include current-integration or “coulomb counting”, which is one of the most common methods used to estimate the SOC. Although current-integration method can be implemented in real-time and can be very accurate, it has one major limitation due to its open-loop nature. That is, with a very small error in the current measurement, the estimated SOC can easily diverge from the true SOC value after few cycles due to accumulated error in the current measurement unless a reset mechanism is periodically used [4] [5] . Another direct method that is used to estimate the SOC is the “open-circuit voltage”. In this method, the voltage is measured continuously and the corresponding SOC is obtained from a lookup table. This method requires a very low (dis)charge current for accurate results, which can be a good choice for some applications that draw very small currents from the battery. Other methods such as the “discharge test” is not practical because it wastes the energy stored in the battery and cannot be used in real-time. In addition, the “impedance spectroscopy” method requires an AC current injection into the cell to measure its impedance and is hard (if not impossible) to achieve dynamically [6] [7].

On the other hand, there are several indirect methods that are used to estimate the SOC. Those methods can be very accurate and reliable in general. Among those methods are extended Kalman filter (EKF) and artificial neural network (ANN). EKF methods employ advanced battery cell models and require a relatively high computation capability. In contrast, ANN methods don’t rely on any electrical, physical, chemical, or thermal model and they require less computation capability compared to EKF methods. However, ANN methods require huge data to train the network in order to calculate the weight functions for the network [4] [8]. In fact, although many work has been reported on EKF ([3] [5] [9] - [14]) and ANN ([15] - [23]) SOC estimators, no evaluation nor comparison on those methods has been reported. This paper is an attempt to evaluate the performance of those methods in estimating the SOC through some experimental battery data. Through this paper, the strengths, drawbacks, and capabilities of both EKF- and ANN-based methods in estimating the SOC will be studied, evaluated and compared.

2. BATTERY DATA

To evaluate and compare the capabilities of the presented methods, some battery cell data were gathered from a commercial 1.1 Ah, 3.6 V Li-ion cell. To ensure that the gathered data represent the steady-state and transient battery responses, a pulse-discharge test was performed over the entire capacity range of the cell (from 100% to 0% SOC). In this test, the reference (true) SOC value is obtained by integrating the current with time (coulomb- counting). The battery was fully charged using constant current/constant voltage (CC/CV) profile, then it was discharged using a 1-Amp pulsed current. The pulse and rest durations were set at 6 and 4 minutes, respectively. Since the test comprises only one cycle, the SOC calculated using coulomb-counting is believed to be 100% accurate and thus it will be used later as a reference. The test was performed at room temperature. Since the SOC is strongly dependent on temperature, the test must be repeated at all possible temperatures to evaluate the methods at any temperature. However, the temperature effect is beyond the scope of this paper and thus the methods were evaluated only at room temperature. Also, the calculated SOC in this test and the estimated SOC in the next two sections are found assuming a constant battery total capacity. In order to further generalize the analysis, the test must be performed at different battery conditions, i.e. different battery ageing levels, which also was not considered in this paper. The voltage and SOC are shown in Figure 1.

3. EXTENDED KALMAN FILTERS (EKF)

3.1. Overview

Kalman filter (KF) is an established technique that is used as an optimal estimator to estimate the inner state of a dynamic linear system. For nonlinear systems, a linearized version called extended Kalman filter or EKF is rather used to estimate a system's state. Basically, KF or EKF is a recursive set of equations that consists of two steps: prediction and correction. In the prediction step, the state of the system is estimated using the previous available measurement, while in the correction step, the estimated state is updated using the current measurement. Figure 2 shows a block diagram of a Kalman filtering process.

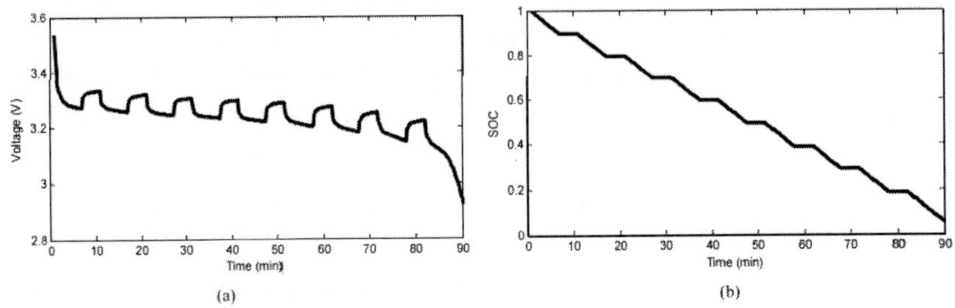

Figure 1. Pulse discharge test: (a) Voltage; and (b) SOC.

3.2. EKF Formulation

If the current is assumed to be positive during charging and negative during discharging, then, the equation of the state-of-charge, $z(t)$, is given in (1).

$$z(t) = z(0) + \eta \int_0^t \frac{i(\tau)}{C_n} d\tau \tag{1}$$

where $z(0)$ is the initial SOC, η is the charging or discharging efficiency, and C_n is the nominal capacity,

which is defined as the number of ampere-hours obtained from a fully charged battery at a $C/30$ rate at room temperature. In a recursive discrete-time form, the last equation is expressed as in (2).

$$z_k = z_{k-1} + \left(\frac{\eta \Delta t}{C_n}\right) i_{k-1} \tag{2}$$

To estimate the SOC, an EKF algorithm is constructed as given in the following equations [24].

- Firstly, the model is expressed in state-space representation

$$z_k = f_{k-1}(z_{k-1}, i_{k-1}, \omega_{k-1}) = z_{k-1} + \left(\frac{\eta \Delta t}{C_n}\right) i_{k-1} + \omega_{k-1},$$
$$y_k = h_k(z_k, i_k, v_k) = h_k(z_k, i_k) + v_k,$$
$$\omega_k \sim (0, Q_k),$$
$$v_k \sim (0, R_k)$$

where k is the time index, $h_k(\cdot)$ is the output equation of the model, ω_k is a discrete-time white noise

(process uncertainty) with a covariance matrix Q_k, and v_k is a discrete-time white noise (measurement uncertainty) with covariance matrix R_k.

- Secondly, the filter is initialized

$$\hat{z}_0^+ = E[z_0] = z_0$$

$$P_0^+ = E\left[(z_0 - \hat{z}_0^+)(z_0 - \hat{z}_0^+)^{\mathrm{T}}\right] = P_{z_0}$$

where P is the estimation error covariance matrix.

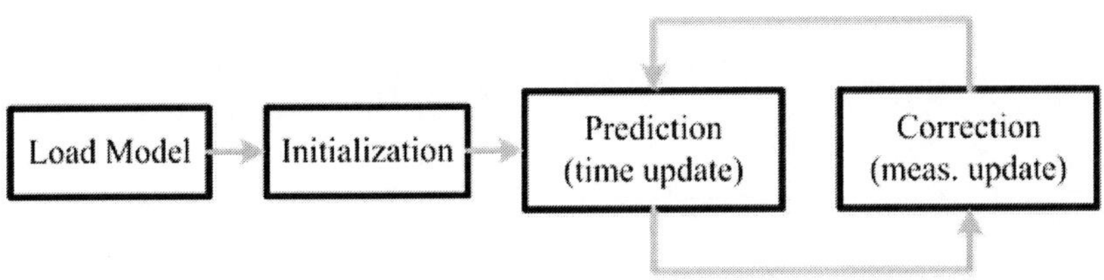

Figure 2. Kalman filtering process.

- Thirdly, for $k = 1, 2, \cdots$, the following steps are performed:

1) Compute the following partial derivative matrices:

$$F_{k-1} = \left.\frac{\partial f_{k-1}}{\partial z}\right|_{\hat{z}_{k-1}^{+}}, \quad L_{k-1} = \left.\frac{\partial f_{k-1}}{\partial \omega}\right|_{\hat{z}_{k-1}^{+}}$$

2) Perform the time update of the state estimate and estimation error covariance:

$$P_k^{-} = F_{k-1} P_{k-1}^{+} F_{k-1}^{\mathrm{T}} + L_{k-1} Q_{k-1} L_{k-1}^{\mathrm{T}} = P_{k-1}^{+} + Q_{k-1},$$
$$\hat{z}_k^{-} = f_{k-1}\left(\hat{z}_{k-1}^{+}, u_{k-1}, 0\right)$$

3) Compute the following partial derivatives:

$$H_k = \left.\frac{\partial h_k}{\partial z}\right|_{\hat{z}_k^{-}} = \frac{y_k\left(\hat{z}_k^{-}\right) - y_{k-1}\left(\hat{z}_k^{-}\right)}{\hat{z}_k^{-} - \hat{z}_{k-1}^{-}}, \quad M_k = \left.\frac{\partial h_k}{\partial v}\right|_{\hat{z}_k^{-}}$$

4) Measurement update:

$$K_k = P_k^{-} H_k^{\mathrm{T}} \left(H_k P_k^{-} H_k^{\mathrm{T}} + M_k R_k M_k^{\mathrm{T}}\right)^{-1},$$
$$\hat{z}_k^{+} = \hat{z}_k^{-} + K_k\left(y_k - h_k\left(\hat{z}_k^{-}, 0\right)\right),$$
$$P_k^{+} = \left(I - K_k H_k\right) P_k^{-}$$

3.3. Battery Modeling and SOC Estimation

To accurately estimate the SOC using EKF (the battery is a nonlinear time-variant dynamic system), an accurate battery model must be available. Some existing battery models are briefly presented [25].

Shepherd Model:

$$y_k = E_o + Ri_k - \frac{\mu}{z_k} \tag{3}$$

Unnewehr Model:

$$y_k = E_o + Ri_k - \mu z_k \tag{4}$$

Nernst Model:

$$y_k = E_o + Ri_k + \mu_1 \ln(z_k) + \mu_2 \ln(1 - z_k) \tag{5}$$

In equations (3)-(5): k is a time index, y_k is the model voltage, E_o is a dc gain, R is the cell internal resistance, i_k is the cell current (positive for charge and negative for discharge), μ , μ_1 and μ_2 are constants used for curve fitting.

Linear Model: Another battery model is the linear equivalent-circuit model shown in Figure 3. This model simply calculates the battery voltage by adding the voltage drop to the open-circuit voltage (OCV) at a given SOC. The output equation of this model is given in (6).

$$y_k = \mathrm{OCV}(z_k) + Ri_k \tag{6}$$

where $i_k > 0$ for charging and $i_k < 0$ for discharging.

RC Model: The last battery model discussed in this paper is the RC equivalent-circuit model. This model consists of an RC network as well as an internal resistance as shown in Figure 4. The RC network is added to implement the time constant during transients, which is equivalent in practice to the cell double layer and diffusion capacitance. The equation for this model is given in (7).

$$y_k = \mathrm{OCV}(z_k) + Ri_k + V_c \exp\left(\frac{-t}{R_c C}\right) \tag{7}$$

where, again, $i_k > 0$ for charging and $i_k < 0$ for discharging.

In (7), t is the transient time when the cell turns from the charge or discharge state to the idle state and vice versa. For both the linear and RC models, the relationship between the SOC and OCV is found by charging and discharging the cell at a very low current while monitoring the SOC. The OCV-SOC pairs are stored in a lookup table. The exact value of the models' parameters in equations (3)-(7) that give the least square error can be found using least square estimation [3] . For the RC model, the internal resistance value is the same as that for the linear model. However, the value of the capacitance cannot be found using least square estimation because the relationship between the output voltage and the parameters of the shunt branch (C and R_c) is nonlinear. Rather, the values of C and R_c were found analytically using trial and error from the pulse discharge test. The model voltage versus the true voltage of the presented five models as well as the results of estimating the SOC using EKF versus the true SOC are shown in Figure 5.

4. ARTIFICIAL NEURAL NETWORKS (ANN)

4.1. Overview

ANN is a subfield of artificial intelligence. It is a mathematical tool that is inspired by the way the human brain processes information [8] . ANNs consist of a number of layers each of which consists of a number of processing units called "neurons" as the basic unit. The neurons are interconnected through weight functions as shown in Figure 6(a). Figure 6(b) shows a block diagram of an adaptive ANNs operation.

ANNs are highly stable and robust due to their massive parallel structure. ANNs consist of inputs and outputs and are made of neurons interconnected with each other. The weights in Figure 7 are determined during the training phase and their values determine how the neural network will respond to a particular time series of inputs. Thus, training ANNs is the most critical phase that determines how accurate those ANNs are. One thing to mention is that ANNs can have an open-loop (feed-forward) or closed-loop (feed-back) structure. Another important feature of ANNs is their ability to adapt to changes (this kind of neural networks is called adaptive artificial neural network). The weights in adaptive ANNs continuously adapt, in real-time, to changes that occurs after training, while non-adaptive ANNs have fixed weights that are determined in the training phase.

The weights of an ANN are calculated in the training phase by minimizing the loss function (usually a quadratic function of the output error). One common technique that is widely used in training ANNs is the back-propagation, which is a supervised learning method that is commonly used in the training phase to calculate the weights of a neural network. Back-propagation uses a steepest-descent technique based on the computation of the gradient of the loss function with respect to the network parameters [26].

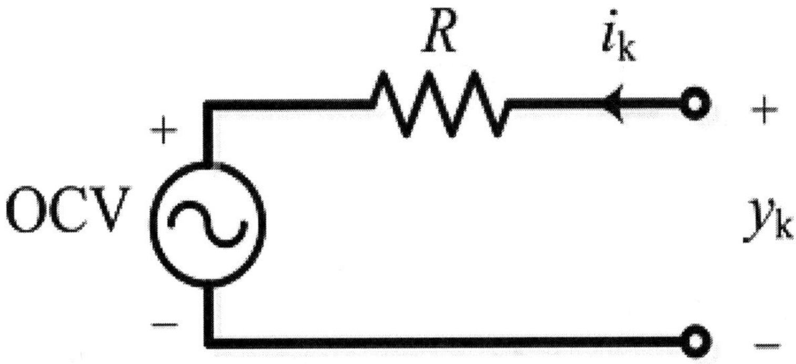

Figure 3. Linear equivalent-circuit model.

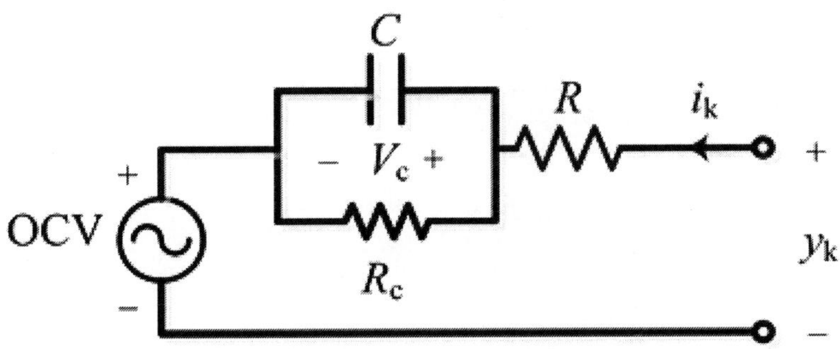

Figure 4. RC equivalent-circuit model.

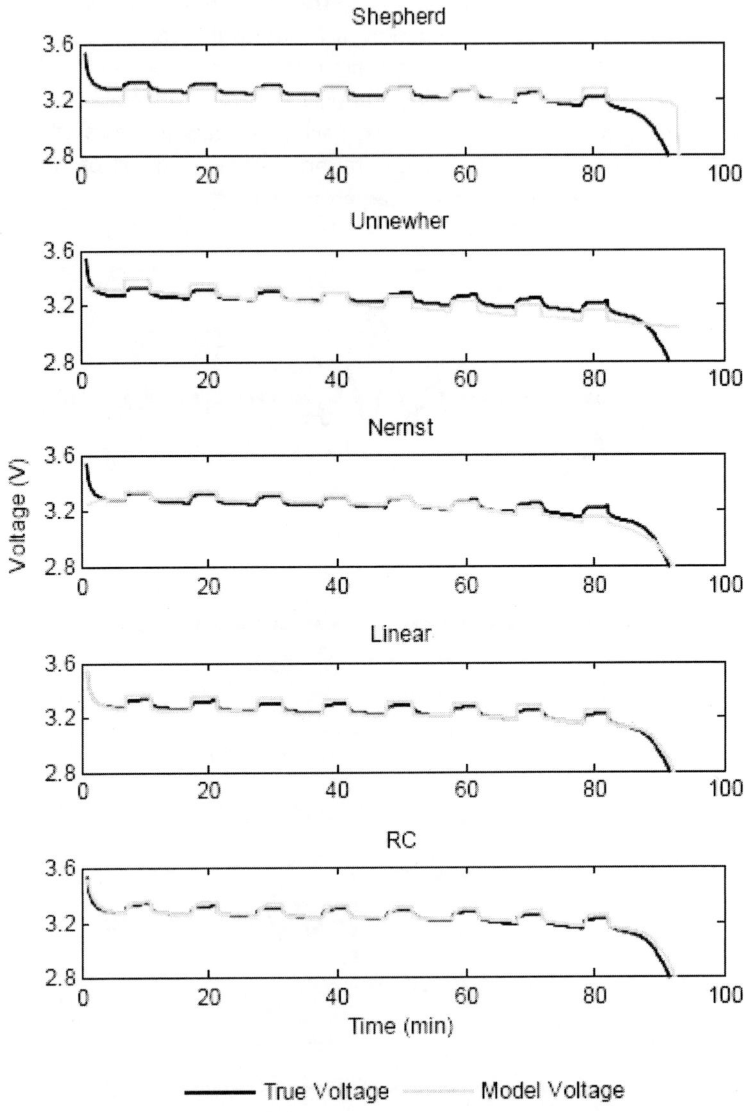
Shepherd
Unnewher
Nernst
Linear
RC
Voltage (V)
Time (min)
True Voltage
Model Voltage

(a)

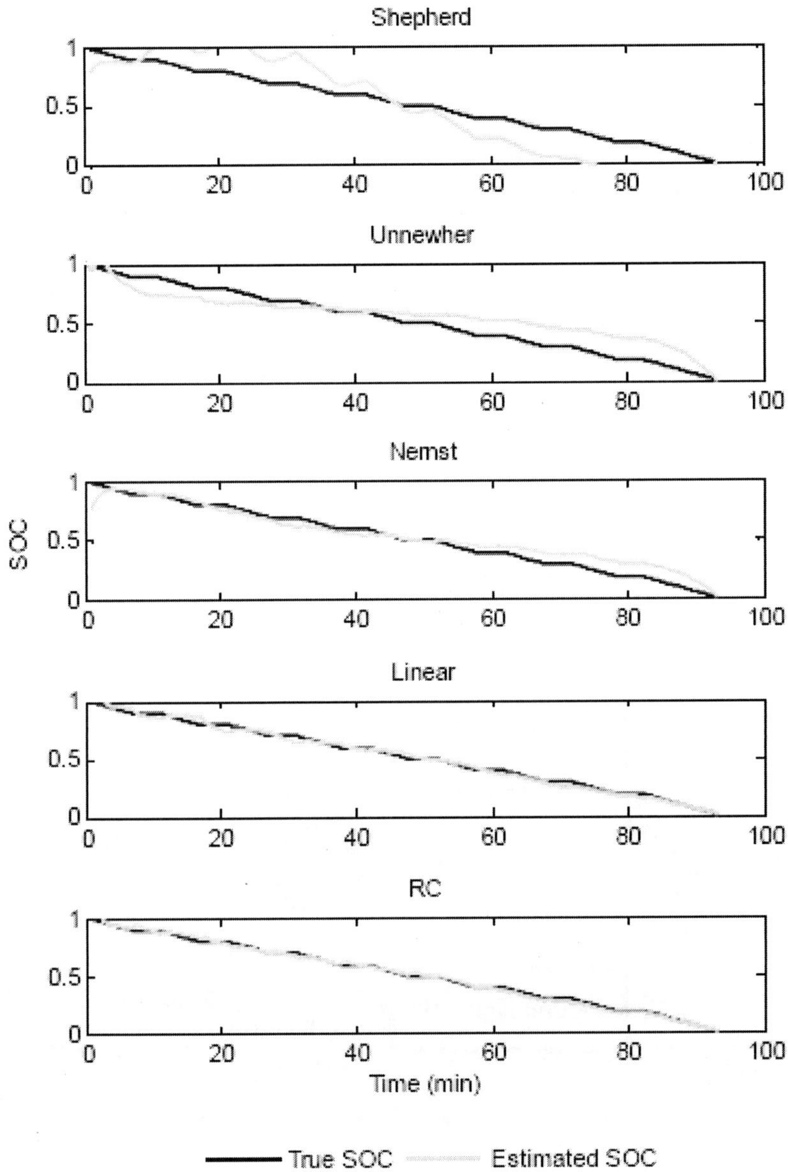

(b)

Figure 5. (a) Results of voltage prediction for the presented models; (b) Results of estimating the SOC using EKF for the presented models.

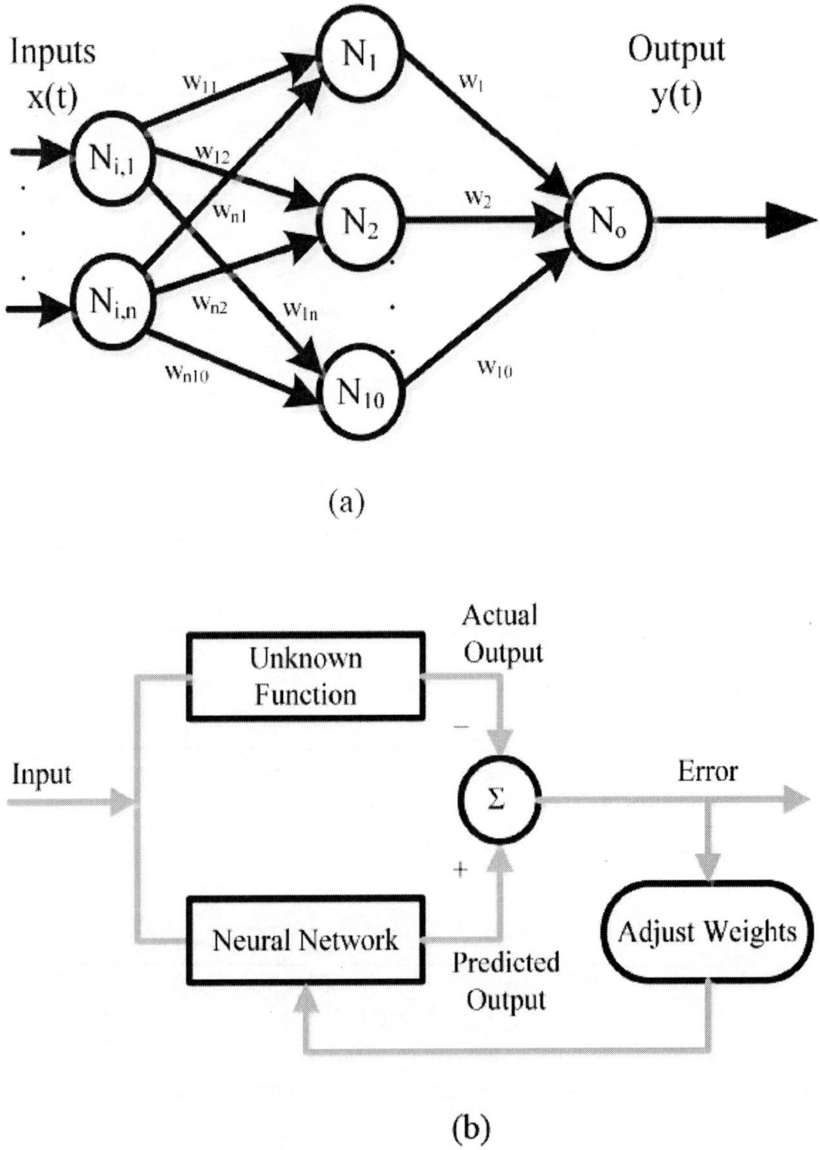

Figure 6. (a) A two-layer feed-forward ANN with 10 hidden neurons, n input neurons and 1 output neuron; (b) Operational block diagram of ANN.

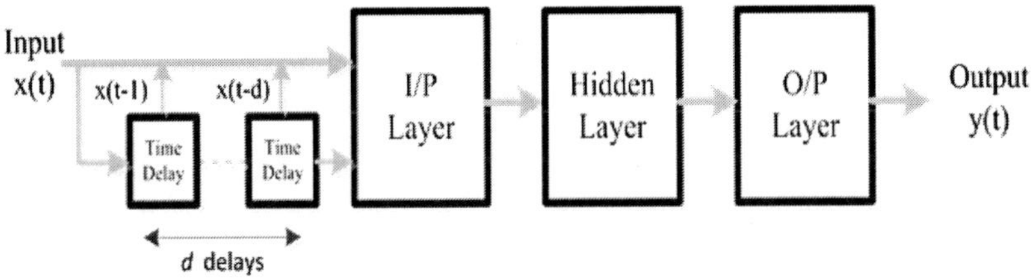

Figure 7. A non-linear input-output (NIO) feed-forward ANN with d time delays.

4.2. SOC Estimation

To evaluate the performance of ANNs in estimating the SOC, two common network architectures were selected: the first is nonlinear input-output (NIO) feed-forward network, and the second is nonlinear autoregressive with exogenous input (NARX) feed-back network. For both networks, Levenberg-Marquardt back-propagation algorithm was used in training.

NIO feed-forward network predicts one time series given the current and past values of an input time series as shown in Figure 7. For the purpose of estimating the SOC, the input time series x(t) consists of the battery terminal voltage and current, where the output or target is the SOC.

The output equation of this network is given in (8).

$$y(t) = f\left(x(t-1),\cdots,x(t-d)\right) \tag{8}$$

NARX feed-back network, on the other hand, predicts one time series given the current and past values of an input time series as well as the past and current values of another time series, called external or exogenous time series (output). A block diagram of this network is shown on Figure 8.

The output equation of this network is given in (9).

$$y(t) = f\left(x(t-1),\cdots,x(t-d),y_2(t-1),\cdots,y(t-d)\right) \tag{9}$$

In (8) and (9): $f(.)$ represents the unknown function in Figure 8, $x(t)$ indicates the input time series that consists of the cell's terminal voltage and current, $y(t)$ is the output time series which is the actual SOC, and d is a delay parameter $(d=1,2,3,\cdots)$.

For both networks, the delay parameter indicates the delay in the input/output time series and it can be selected by trial and error until the desired accuracy is obtained. The number of neurons in the hidden layer was set at 10. The value of the delay parameter and the number of neurons in the hidden layer were purposefully varied in order to find the effect of the selected values on the network performance.

The results of estimating the SOC for both the NIO and NARX models are shown in Figure 9. The value of the delay parameter d was values of 2, 4, 6, 8 and 10.

5. COMPARATIVE SUMMARY OF RESULTS

In order to obtain practical results, a zero-mean process noise with an absolute peak value of 10 mA and a zero-mean measurement noise with an absolute peak value of 10 mV were added to the current and voltage readings. The results in Figure 5 and Figure 9 are obtained after the process and measurement were contaminated. Both the process and measurement noises are independent and Gaussian. The presented methods are evaluated and compared in the rest of this section.

It is obvious in Figure 5(b) that the accuracy of the EKF algorithms is totally dependent on the accuracy of the model. Actually, this total dependence of EKFs on the adopted model is problematic and can result in inaccurate and instable performance when estimating the battery SOC. In other words, a small mismatch between the model voltage and the true voltage can lead the estimated SOC to diverge from the true SOC. In the presented EKF algorithms, Nernst model performed better than Shepherd and Un-newehr models because it employs two correction factors. The linear and RC equivalent circuit models performed much better than Shepherd, Un-newehr and Nernst models because they are more advanced since they use an OCV-SOC lookup table. The RC model outperformed all the other models because it handles both the internal voltage drop and transient (capacitance) voltage of the battery.

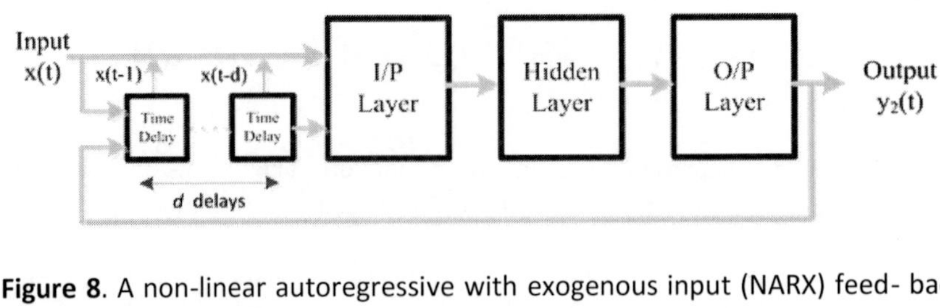

Figure 8. A non-linear autoregressive with exogenous input (NARX) feed- back ANN with d time delays.

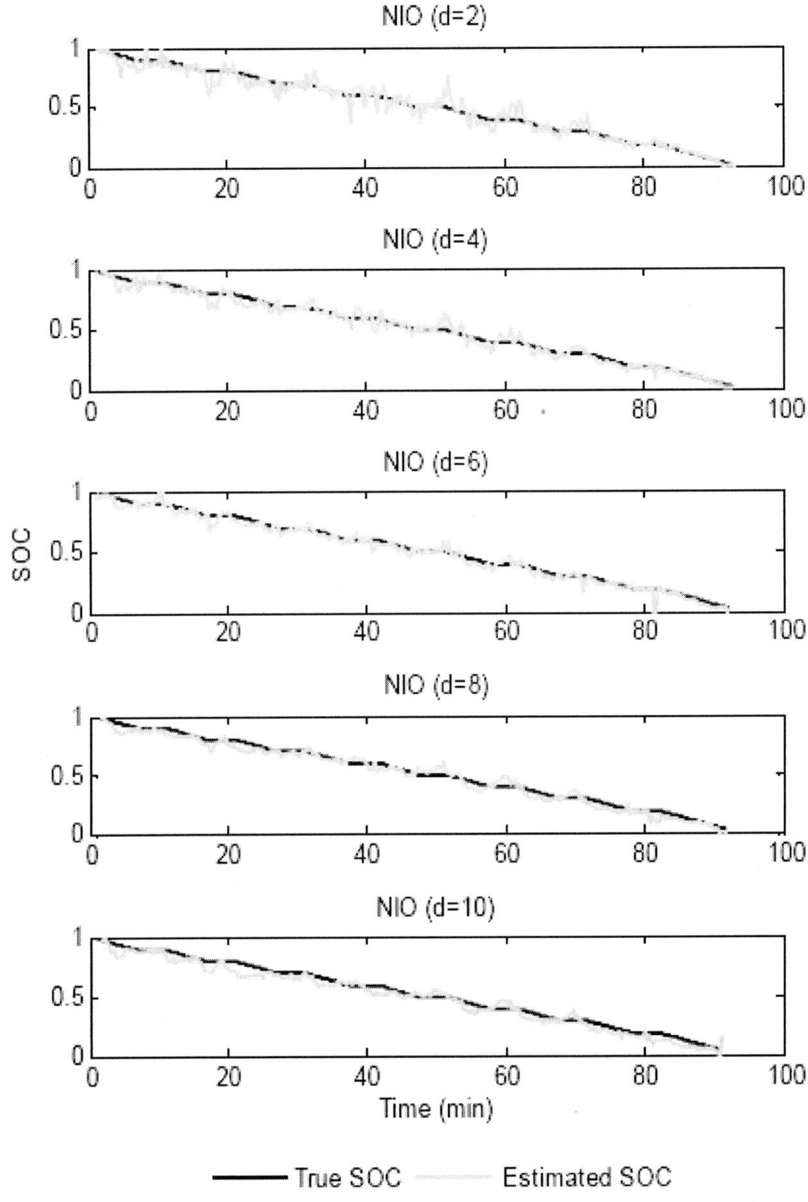
NIO (d=2)
NIO (d=4)
NIO (d=6)
NIO (d=8)
NIO (d=10)
SOC
Time (min)
True SOC
Estimated SOC

(a)

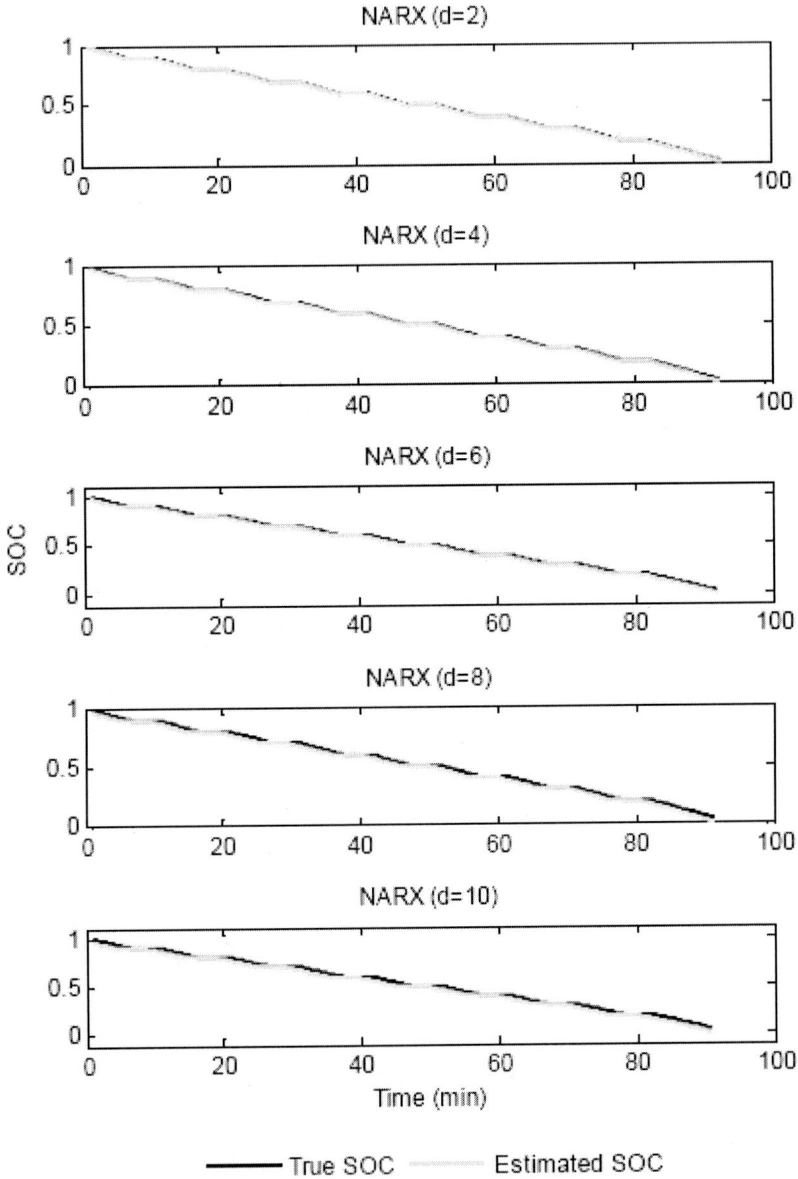

(b)

Figure 9. Results of estimating the SOC with delay values equal 2, 4, 6, 8 and 10 using (a) NIO feed-forwardANN, and (b) NARX feed-back ANN.

In fact, one advantage of EKFs is that they can still be accurate even if the initial guess was incorrect. To verify that, the initial SOC was purposefully changed from 100% (true value) to 80% (wrong value), and it was noticed that every time the algorithm converged to the true value after a relatively short time (within 5 minutes). Another advantage of EKFs is that they can also provide accurate performance even when the process and measurement have noise. To verify this, a white noise (Gaussian) was added to both the current and voltage. All the EKF algorithms, regardless what battery model they used, were able to achieve almost the same results before the process and measurement were contaminated.

In practice, however, the process and measurement noise is not necessarily white. If so, conventional EKFs will not perform well. Thus, EKFs can be a good option only and only if an accurate battery and noise models are used; otherwise, EKFs will not be accurate and will be prone to stability issues.

Table 1. Evaluation of reviewed methods in terms of RMSE and linear regression.

Model	RMSE	Regression
Shepherd	0.1739	0.95478
Unnewehr	0.1178	0.95155
Nernst	0.0672	0.98321
Linear	0.0284	0.99530
RC	0.0202	0.99788
NIO ($d = 2$)	0.0579	0.97694
NIO ($d = 4$)	0.0433	0.98742
NIO ($d = 6$)	0.0379	0.99176
NIO ($d = 8$)	0.0421	0.98995
NIO ($d = 10$)	0.5889	0.98860
NARX ($d = 2$)	0.0074	0.99990
NARX ($d = 4$)	0.0128	0.99961
NARX ($d = 6$)	0.0193	0.99920
NARX ($d = 8$)	0.0253	0.99867
NARX ($d = 10$)	0.0311	0.99807

On the other hand, the presented ANNs showed different performances. NARX networks obviously outperformed NIO networks due to their feedback loop. In the presented NARX network, the feedback loop was used in the training phase to calculate the weight functions. The feedback resulted in a more accurate performance as can be seen by comparing the NARX and NIO models. However, although the calculated weight functions for both the NIO and NARX models are optimal in the sense that they gave the best possible performance, the weight functions must be able to adapt to changes in the ambient conditions or battery parameters.

One remark on the presented ANNs is the relationship between the delay parameter d and the accuracy of the model. NIO networks showed the best performance at $d = 6$, while for the NARX networks, the mismatch between the true and estimated SOC increased as d increased. Nonetheless, NIO networks can still be a good choice when no output data is available. It was shown that as d increases in the NIO network, the noise in the estimated SOC decreases, while for the NARX network, the were no noise in all cases, which is another advantage of the feed-back loop.

Another remark on the ANNs models is that there was almost no relationship between the network accuracy and the number of neurons in the hidden layer. It was noticed that when the number of neurons in the hidden layer is less than 5, the accuracy is slightly lower than when the number of neurons is above 5. However, the performance of the network didn't change at all when the number of neurons in the hidden layer was changed randomly between 5 and 100 for both NIO and NARX networks.

To quantitatively compare the presented methods, the root mean square error (RMSE) and regression were calculated and summarized in Table 1.

The regression in Table 1 was found using MATLAB function (plotregression), where the RMSE was found using equation (10).

$$\mathrm{RMSE} = \sqrt{\frac{1}{N}\sum_{k=1}^{N}\left(z_{\mathrm{true}} - z_{\mathrm{est}}\right)^2} \tag{10}$$

where z_{true} and z_{est} are the true and estimated SOC, and N is the number of estimations.

6. CONCLUSION

Two advanced SOC estimation methods were analyzed, compared and evaluated. The two methods, EKFs and ANNs, can perform well under certain conditions. That is, EKFs can perform well if an accurate battery and noise models are used, where ANNs can perform well if they were trained with an artificial experience (training) that is very close or identical to the real experience. However, even with ideal battery model or training, the actual EKF or ANN SOC estimator will not be ideal due to the variation between battery cells in their internal parameters even if they were from the same manufacturer. Other factors that must be taken into account when designing an SOC estimator are the dependence of the battery on temperature and the change in the battery parameters (i.e. capacity, internal resistance, etc.) through the battery lifetime. Thus, adaptability is a key issue for highly reliable SOC estimators through the battery lifetime.

ACKNOWLEDGEMENTS

This research was supported in part by the United Arab Emirates University under startup research grant # G00001287.

REFERENCES

1. Cope, R.C. and Podrazhansky, Y. (1999) The Art of Battery Charging. The 14th Annual IEEE Battery Conference on Applications and Advances, Long Beach, 12-15 January 1999, 233-235.
2. Hussein, A.A. and Batarseh, I. (2011) A Review of Charging Algorithms for Nickel and Lithium Battery Chargers. IEEE Transactions on Vehicular Technology, 60, 830-838.
3. Hussein, A.A. and Batarseh, I. (2011) State-of-Charge Estimation for a Single Lithium Battery Cell Using Extended Kalman Filter. IEEE Power and Energy Society General Meeting, 24-29 July 2011, San Diego, 1-5.
4. Hussein, A.A. (2013) Capacity Fade Estimation in Electric Vehicles Li-Ion Batteries Using Artificial Neural Networks. 2013 ECCE Conference, Denver, 15-19 September 2013, 677-681.
5. Plett, G.L. (2004) Extended Kalman Filtering for Battery Management Systems of LiPB-Based HEV Battery Packs: Part 3. State and Parameter Estimation. Journal of Power Sources, 134, 277-292.
6. Piller, S., Perrin, M. and Jossen, A. (2001) Methods for State-of-Charge Determination and Their Applications. Journal of Power Sources, 96, 113-120.
7. Hussein, A.A., Kutkut, N., Shen, J. and Batarseh, I. (2012) Distributed Battery Micro-Storage Systems Design and Operation in a Deregulated Electricity Market. IEEE Transactions on Sustainable Energy, 3, 545-556.

8. Haykin, S. (1999) Neural Networks and Learning Machines. 3rd Edition, 1-46.

9. He, H., Xiong, R., Zhang, X. and Sun, F. (2011) State-of-Charge Estimation of the Lithium-Ion Battery Using an Adaptive Extended Kalman Filter Based on an Improved Thevenin Model. IEEE Transactions on Vehicular Technology, 60, 1461-1469.

10. Zhang, F., Guangjun, L. and Fang, L. (2008) A Battery State of Charge Estimation Method with Extended Kalman Filter. IEEE/ASME International Conference on Advanced Intelligent Mechatronics, Xian, 2-5 July 2008, 1008-1013.

11. Qiu, S., Chen, Z., Masrur, M.A. and Murphey, Y.L. (2011) Battery Hysteresis Modeling for State of Charge Estimation Based on Extended Kalman Filter. Proceedings of the 6th IEEE Conference on Industrial Electronics and Applications, Beijing, 21-23 June 2011, 184-189.

12. Domenico, D.D., Fiengo, G. and Stefanopoulou, A. (2008) Lithium-Ion Battery State of Charge Estimation with a Kalman Filter Based on a Electrochemical Model. Proceedings of the IEEE International Conference on Control Applications, San Antonio, 3-5 September 2008, 702-707.

13. Yan, W., Ming, Y.T. and Jie, L.B. (2008) Lead-Acid Power Battery Management System Basing on Kalman Filtering. Proceedings of the IEEE Vehicle Power and Propulsion Conference, Harbin, 3-5 September 2008, 1-6.

14. Windarko, N.A., Choi, J. and Chung, G.B. (2011) SOC Estimation of LiPB Batteries Using Extended Kalman Filter Based on High Accuracy Electrical Model. Proceedings of the 8th IEEE International Conference on Power Electronics and ECCE Asia, Jeju, 30 May-3 June 2011, 2015-2022.

15. Chan, C.C., Lo, E.W.C. and Weixiang, S. (2000) The Available Capacity Computation Model Based on Artificial Neural Network for Lead-Acid Batteries in Electric Vehicles. Journal of Power Sources, 87, 201-204.

16. Shi, P., Bu, C. and Zhao, Y. (2005) The ANN Models for SOC/BRC Estimation of Li-Ion Battery. Proceedings of the IEEE International Conference on Information Acquisition, Hong Kong and Macau, 27 June-3 July 2005.

17. Shen, W.X., Chan, C.C., Lo, E.W.C. and Chau, K.T. (2002) A New Battery Available Capacity Indicator for Electric Vehicles Using Neural Network. Energy Conversion and Management, 43, 817-826.

18. Yamazaki, T., Sakurai, K. and Muramoto, K. (1998) Estimation of the Residual Capacity of Sealed Lead-Acid Batteries by Neural Network.

Proceedings of the 20th International Telecommunications Energy Conference, San Francisco, 4-8 October 1998, 210-214.

19. Affanni, A., Bellini, A., Concari, C. and Franceschini, G. (2003) EV Battery State of Charge: Neural Network Based Estimation. Proceedings of the IEEE International Electric Machines and Drives Conference, Madison, 1-4 June 2003, 684-688.

20. Shen, W.X. (2007) State of Available Capacity Estimation for Lead-Acid Batteries in Electric Vehicles Using Neural Network. Energy Conversion and Management, 48, 433-442.

21. Shen, W.X., Chau, K.T., Chan, C.C. and Lo, E.W.C. (2005) Neural Network-Based Residual Capacity Indicator for Nickel-Metal Hydride Batteries in Electric Vehicles. IEEE Transactions on Vehicular Technology, 54, 1705-1712.

22. Cai, C.H., Du, D., Liu, Z.Y. and Zhang, H. (2002) Artificial Neural Network in Estimation of Battery State of Charge (SOC) with Nonconventional Input Variables Selected by Correlation Analysis. Proceedings of the International Conference on Machine Learning and Cybernetics, Beijing, 4-5 November 2002, 1619-1625.

23. Chen, Z., Qiu, S., Masrur, M.A. and Murphey, Y.L. (2011) Battery State of Charge Estimation Based on a Combined Model of Extended Kalman Filter and Neural Networks. Proceedings of the 2011 International Joint Conference on Neural Networks (IJCNN), San Jose, 31 July-5 August 2011, 2156-2163.

24. Simon, D. (2006) Optimal State Estimation. Wiley-Interscience, Hoboken, 407-409.

25. Hussein, A.A. and Batarseh, I. (2011) An Overview of Generic Battery Models. Proceedings of the IEEE Power and Energy Society General Meeting, San Diego, 24-29 July 2011, 1-6.

26. Hippert, H.S., Pedreira, C.E. and Souza, R.C. (2001) Neural Networks for Short-Term Load Forecasting: A Review and Evaluation. IEEE Transactions on Power Systems, 16, 44-55.

Chapter 5

A Novel Grouping Method for Lithium Iron Phosphate Batteries Based on a Fractional Joint Kalman Filter and a New Modified K-Means Clustering Algorithm

Xiaoyu Li, Kai Song, Guo Wei*, Rengui Lu and Chunbo Zhu*

School of Electrical Engineering and Automation, Harbin Institute of Technology, Harbin 150001, China

ABSTRACT

This paper presents a novel grouping method for lithium iron phosphate batteries. In this method, a simplified electrochemical impedance spectroscopy (EIS) model is utilized to describe the battery characteristics. Dynamic stress test (DST) and fractional joint Kalman filter (FJKF) are used to extract battery model parameters. In order to realize equal-number grouping of batteries, a new modified K-means clustering algorithm is proposed. Two rules are designed to equalize the numbers of elements in each group and exchange samples among groups. In this paper, the principles of battery model selection, physical meaning and identification method of model parameters, data preprocessing and equal-number clustering method for battery grouping are comprehensively described. Additionally, experiments for battery grouping and method validation are designed. This method is meaningful to application involving the grouping of fresh batteries for electric vehicles (EVs) and screening of aged batteries for recycling.

Keywords: Battery grouping; Fractional joint Kalman Filter; equal-number; Modified K-means clustering

1. INTRODUCTION

With the development of EVs, battery technology has drawn more and more attention worldwide. Battery packs are core components of EVs; they are composed of hundreds or thousands of small cells joined by series-parallel connections. Owing to the subtle differences in the battery production process, the electric characteristics of the batteries are slightly different. The current flowing through the batteries will thus be inconsistent and the temperature inside the battery pack will be uneven when the inconsistent batteries are connected in parallel. In this case, battery use will accelerate their decay and some safety problems will occur [1]. In series connection, the potential of most batteries in the battery pack will not exhaust. Moreover, battery state monitoring and management, such as state of charge (SoC), state of health (SoH) estimation [2,3,4] and state of peak power (SoP) prediction [5,6], will become more and more difficult. Therefore, measures are necessary to guarantee the characteristics of the grouped batteries are as similar as possible.

Numerous battery grouping methods have been reported in previous studies. Kim *et al.* proposed a battery voltage and SoC consistency screening method based on matching battery DC resistance [7,8]. Schneider *et al.* used a sorting method for aged batteries [9]. In that paper, the sorting steps for these batteries are designed. Fang *et al.* proposed a method for battery classification based on the thermal behavior during the charging process [10]. In the method, differences in battery surface temperature are used to evaluate the consistency of batteries. At present, battery capacity, AC resistance, electrochemical impedance spectroscopy (EIS), voltage curve, battery model parameter, charge and discharge thermal behavior are the commonly used parameters to evaluate consistency.

In the authors' previous work [11], the reliability and effectiveness of these evaluation bases are comprehensively analyzed. The main arguments are summarized as follows: AC resistance represents only a small part of the impedance characteristics of the battery, and this parameter cannot reflect the dynamic characteristics of the battery, since the testing frequency (about 1 kHz) is much higher than the main frequency component of the driving conditions. Battery charge and discharge voltage curves can reflect the voltage consistency under the test conditions, however, the reliability of any conformance assessment will be decreased if the working conditions change. Additionally, the voltage curve is just an external manifestation of the battery, and the internal characteristics' consistency cannot be guaranteed because of the complex relationship between voltage curve and internal material properties. The surface temperature of a battery is determined by the entropy heat, resistance heat and surface heat dissipation. The parameters can reflect the battery performance, but they are susceptible to the external environment. EIS or battery model parameters can better reflect the dynamic characteristics of a battery. Capacity consistence determines that the performance of batteries in the battery pack cannot reach their full potential. However, battery capacity is obtained by constant current charge and discharge tests, EIS is measured by sweep frequency impedance tests and battery model parameters are obtained by intermittent constant current and pulse current tests. Considerable time and special testing instruments are required for these three tests.

The existing battery clustering algorithms include the self-organizing feature map algorithm (SOM) [10], fuzzy c-means (FCM) clustering algorithm, *etc.* [12,13,14]. These earlier works focus more on the similarity of batteries' characteristics. However, the number of elements in each group after grouping is different and not equal to the design number of batteries in each pack.

After review of the problems of the battery consistency evaluation parameter extraction methods and battery clustering algorithms, this paper presents a novel battery grouping method. This method can be divided into two parts: the first part is the battery characteristic parameter extraction, which is based on a simplified EIS model, fractional joint Kalman filter algorithm and DST—the details are described in Section 2. The second part is a new modified K-means clustering algorithm especially proposed for battery equal-number grouping—this part are presented inSection 3. The proposed method is established based on a short time charge and discharge experiment. The characteristic parameters including open circuit voltage (OCV), simplified EIS model parameters and battery capacity can be obtained. All these parameters have relatively clear physical meanings. The clustering algorithm can divide the batteries into equal-number groups according to the similarity of consistent evaluation parameters and the number of elements designed by engineers in each group.

2. BATTERY PARAMETER EXTRACTION

2.1. Battery Model Selection

2.1.1. Comparison of Battery Models

It is essential to select a suitable model for quantifying battery characteristics. Recent battery models mainly involve equivalent circuit models based on the external electric characteristics of battery charge and discharge (ECMs for short in this paper) or electrochemical impedance spectroscopy tests (EIS models), electrochemical reaction mechanism models (ERM models), *etc.* In ECMs, resistance, capacitor and diodes are commonly used. Their structures are simple and easy to calculate. These models are widely used in battery state monitoring and management applications [5,15,16,17,18,19]. The typical equivalent circuit models include the R_{int} model, first order RC model, PNGV model, *etc.* Among them, the first order RC model is the most widely used because of its simple structure and high accuracy. This model is shown in Figure 1a.

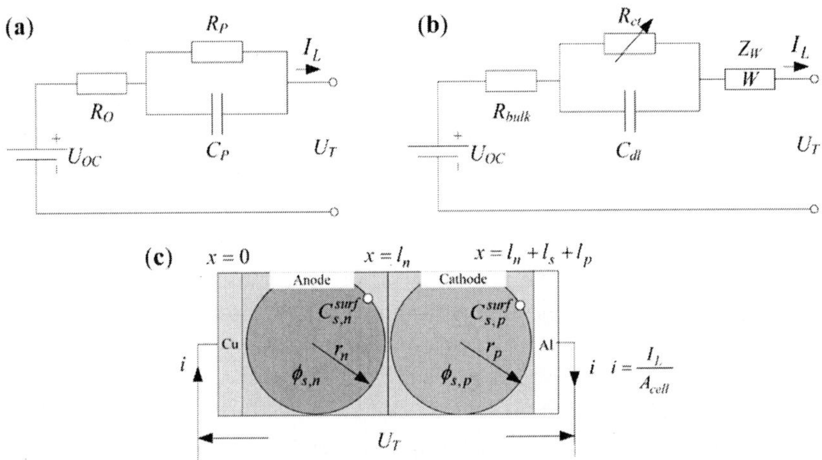

Figure 1. (**a**) First-order RC model; (**b**) a simple EIS model; (**c**) single particle model.

EIS is usually used to study the electrochemical process of the battery electrode/electrolyte interface and in the analysis of the insertion and extraction process of the battery electrode active material. Recently, the EIS technique has had a wide range of applications in battery charging and discharging dynamic modeling. Yoon *et al.* [20] have used impedance data to evaluate the power capacity of batteries. Xu *et al.* [21] have used a lithium-ion battery EIS model and fractional order Kalman filtering algorithm to achieve accurate battery SoC estimation. Waag *et al.* [22] have used an EIS model to analyze the aging state of batteries. Compared with ECMs, EIS models not only reflect the dynamic characteristics of battery more accurately, but also the structure elements have clear physical meanings, therefore these models have attracted more and more attention. Traditional EIS tests are conducted in a laboratory environment. During the process, a small current or voltage is loaded onto the battery and the response is measured. A simple EIS model is shown in Figure 1b.

The behavior inside a battery, including the electrochemical kinetics and charge transfer processes are described in ERM models [23]. The relationship between material properties and a battery's electrical performance are established in these models. In these models, the pseudo two-dimensional model (P2D model) created by Fuller *et al.* [24] and single particle model (SP model) [25] are the most representative ones. Among these, the, SP model is most the simple ERM. The structure of a SP model is shown in Figure 1c.

For battery grouping applications, too many characteristic parameters will lead to an increase in the dimensionality of the battery characteristic vector and battery clustering will be more difficult. If the weight setting of parameters is unreasonable, the accuracy of the clustering results will be reduced. Hence, the battery model structure used for battery grouping should be simple and better reflect the battery dynamics. In the examples described above, EIS models are more suitable for battery grouping due to their fewer parameters and clear physical meaning.

2.1.2. Simplified EIS Model

The EIS test result of a lithium iron phosphate ($LiFePO_4$/graphite) battery and its fitting result based on the simple EIS model are presented in Figure 2.

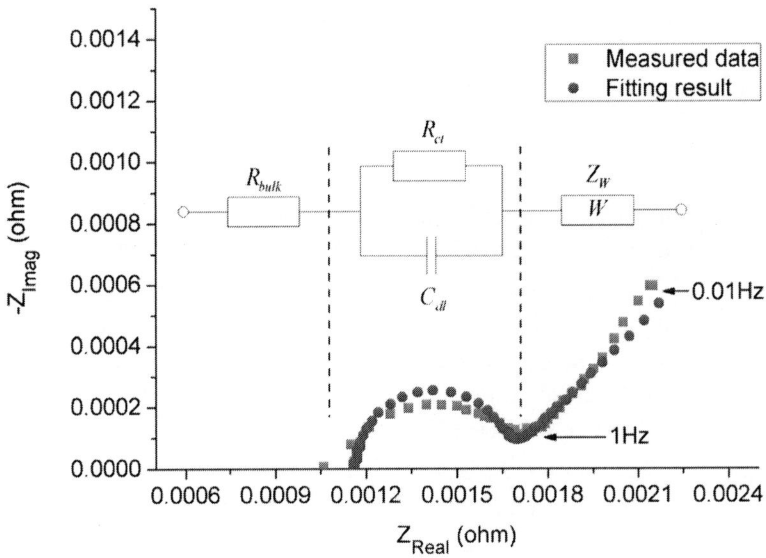

Figure 2. EIS test result and fitting result of the simple EIS model.

The values of R_{ct} and C_{dl} are 0.46 mΩ and 28.59 F, and the time constant τ is 0.01315 s. The time constant of R_{ct} and C_{dl} is very small and barely manifested on the terminal voltage in the case of electric vehicle applications. Hence, the simple EIS model can be further simplified for a $LiFePO_4$/graphite battery.

The impact of double layer capacitance on battery's external characteristics can be ignored. Moreover, R_{ct} is assumed to be constant with current at the room temperature. Thereby, the simplified EIS model shown in Figure 3 is obtained.

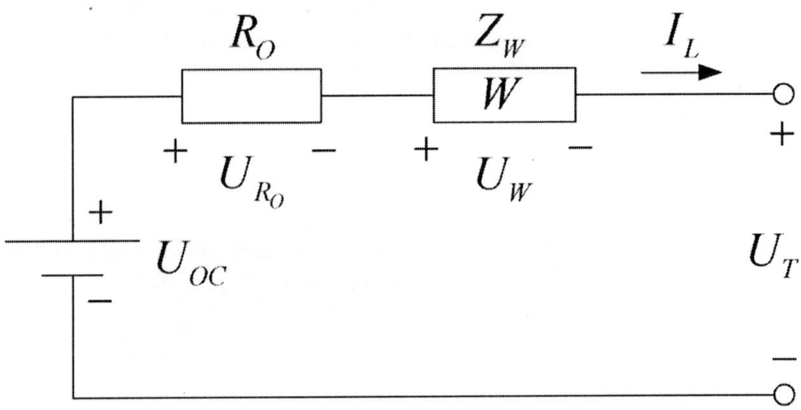

Figure 3. The simplified EIS model.

The model consists of three elements: voltage source U_{OC}, ohmic resistance R_O and Warburg impedance Z_W. U_T and I_L denote the terminal voltage and the total current flowing through the battery, respectively.

Voltage source is used to describe the open-circuit voltage characteristics of the battery. Due to the simplification of the battery model, many chemical reaction kinetics processes and the chemical reaction boundary conditions are ignored. U_{OC} represents the value of the open circuit voltage (*OCV*), and can be expressed by Equation (1) [26]:

$$U_{OC} = E_{OC,p}\left(SoC_p\right) - E_{OC,n}\left(SoC_n\right) \quad (1)$$

where $E_{OC,p}$ and $E_{OC,n}$ are open circuit potentials of positive and negative electrodes. They are functions of the *SoC* which is determined by the average concentration of ions of each electrode. U_{OC} is mainly decided by three parameters: the *SoC*'s start points of positive and negative electrodes and the cycle range of *SoC*.

R_O is a pure resistance element, which mainly reflects the medium-high frequency (typically <1 Hz) impedance characteristic of EIS; the value of this parameter is approximate the sum of the battery bulk resistance R_{bulk}, SEI film resistance R_{set} and electric charge transfer resistance R_{ct} [27], as is shown in Equation (2):

$$\begin{cases} R_O = R_{bulk} + R_{sei} + R_{ct} \\ R_{bulk} = R_{ext} + R^{eff}_{O,\text{electrolyte}} + R^{eff}_{O,solid} \\ R_{sei} = A\dfrac{\eta_{sei}}{I_L} \\ R_{ct} = \dfrac{\eta_{ct}}{I_L}, I_L = A \cdot i_0 \cdot \left(\exp\left(\dfrac{\alpha_a F}{RT}\right)\eta_{ct} - \exp\left(\dfrac{\alpha_c F}{RT}\right)\eta_{ct} \right) \end{cases} \tag{2}$$

where R_{ext} is the connecting resistance, $R^{eff}_{O,\text{electrolyte}}$, electrolyte and $R^{eff}_{O,solid}$, solid stand for the effective part of the electrolyte ohm impedance and solid ohm impedance affecting on the terminal voltage. R_{sei} is mainly determined by the SEI film produced on the surface of negative electrode particles. R_{ct} is decided by Butler-Volmer kinetics.

Z_W is the Warburg impedance, which is usually used to describe the diffusion characteristics. It is a 0.5 order fractional element. U_W is the terminal voltage of this element. Fleischer *et al.* [28] presented the physical meaning of the parameters in Z_W; without considering the boundary conditions, Z_W can be expressed by Equation (3):

$$Z_W = \frac{\mathrm{R}T}{cn^2F^2A} \cdot \frac{1}{\sqrt{jwD}} \tag{3}$$

where, R is the gas constant, *T* is the temperature, *c* represents the molar concentration of active particles, *F* is Faraday's constant, *A* is the active surface area, *D* is the diffusion coefficient of the materials. $X_W = \dfrac{RT}{cn^2F^2AD^{0.5}}$, then Equation (4) can be obtained:

$$Z_W = \frac{X_W}{(jw)^{0.5}} \tag{4}$$

The characteristics of the battery from different aspects are described by the three model parameters. Meanwhile, in the simplified EIS model, charge capacity (Q) is also a crucial parameter for the batteries, and it is meaningful to battery grouping. Four battery performance parameters, Q, U_{OC}, R_O, X_W are used for battery grouping.

2.2. Model Parameter Identification

The simplified EIS model parameter online identification method for EVs is proposed in the authors' previous work [29]. In the method, a model discrete state equation should be established first, and then, model parameters are identified based on the fractional joint Kalman filter algorithm. The details of the two parts are described as follows.

2.2.1. Model State Equation Establishment

The simplified EIS model has been established in Section 2.1. In this section, the state and observation equations are constructed as follows, taking the discharge current value $I_{L,dis}$ as the positive value and data sampling period T_s is one second. In Equation (5), Δ^r is used as a differential operator:

$$\Delta^r = \frac{d^r}{dt^r}, \quad r > 0 \tag{5}$$

where, r is the differential order. When $r = 1$, Δ^1 represents a one order differential operator. As the commonly used capacitor element, if its capacitance is C, the relationship of terminal voltage U_C and the current I_C flowing through this element can be expressed by Equation (6):

$$\Delta^1 U_C = \frac{1}{C} I_C \tag{6}$$

When r is a fractional value, Δ^r represents a fractional differential operator. Similar to the capacitor, the relationship of terminal voltage U_W and the current I_L flowing through the element can be expressed as follows:

$$\Delta^{0.5} U_W = X_W I_L \tag{7}$$

The battery model parameters X_W, U_{OC}, R_O change slightly with *SoC*. Then, the differential equations of state and parameters can be written in the following matrix form:

$$\Delta^{\begin{bmatrix}0.5\\1\\1\\1\end{bmatrix}}\begin{bmatrix}U_W\\X_W\\U_{OC}\\R_O\end{bmatrix}=\begin{bmatrix}0 & I_L & 0 & 0\\0 & 0 & 0 & 0\\0 & 0 & 0 & 0\\0 & 0 & 0 & 0\end{bmatrix}\begin{bmatrix}U_W\\X_W\\U_{OC}\\R_O\end{bmatrix} \tag{8}$$

U_T is treated as the measurement observation parameter. The observation equation is written as below:

$$U_T = U_{OC} - I_L R_O - U_W \tag{9}$$

Equation (8) can be discretized using a bilinear transformation method. Additionally, the system state noise ω and observation noise υ are considered in these equations, where these noises are assumed independent with each other. The discrete form of the state and observation equations is expressed by Equation (10):

$$\begin{cases}\Delta^{\begin{bmatrix}0.5\\1\\1\\1\end{bmatrix}}\begin{bmatrix}U_W\\X_W\\U_{OC}\\R_O\end{bmatrix}_k=\begin{bmatrix}0 & \dfrac{I_{L,k}+I_{L,k-1}}{2} & 0 & 0\\0 & 0 & 0 & 0\\0 & 0 & 0 & 0\\0 & 0 & 0 & 0\end{bmatrix}\begin{bmatrix}U_W\\X_W\\U_{OC}\\R_O\end{bmatrix}_{k-1}+\omega\\ U_{t,k}=\begin{bmatrix}-1 & 0 & 1 & -I_{L,k}\end{bmatrix}\begin{bmatrix}U_W & X_W & U_{oc} & R_o\end{bmatrix}_k^T+\upsilon\end{cases} \tag{10}$$

Based on the definition of the Grünwald-Letnikov fractional differential [21,30], the discrete differential form of a fractional state variable can be written as:

$$\Delta^r x_k = \sum_{j=0}^{k}(-1)^j\binom{r}{j}x_{k-j},\ \binom{r}{j}=\begin{cases}1 & \text{for } j=0\\ r(r-1)\ldots(r-j+1)/j! & \text{for } j>0\end{cases} \tag{11}$$

In this paper, the following definitions are made:

$$\gamma_j = \mathrm{diag}\left[\begin{pmatrix}0.5\\ j\end{pmatrix} \begin{pmatrix}1\\ j\end{pmatrix} \begin{pmatrix}1\\ j\end{pmatrix} \begin{pmatrix}1\\ j\end{pmatrix}\right] \tag{12}$$

$$N = \begin{bmatrix}0.5 & 1 & 1 & 1\end{bmatrix}^T \tag{13}$$

where x in Equation (11) is $[U_W\, X_W\, U_{OC}\, R_O]^T$, then, the discrete differential form of the state and parameters of the simplified EIS model can be obtained:

$$\Delta^N x_k = x_k + \sum_{j=1}^{k} (-1)^j \gamma_j x_{k-j} \tag{14}$$

In Equation (14), the complexity of $\sum_{j=1}^{k} (-1)^j \gamma_j x_{k+1-j}$ is increased with the length of time, so this feature is not suitable for engineering applications. Additionally, the battery diffusion process does not occur infinitely, as it is constrained by the boundary conditions. Thus, k is replaced with variable length L. The value of L is chosen by applying curve fitting to the experimental data. Finally, the state and observation equations can be obtained and are shown in Equation (15):

$$\begin{cases} x_k = \underbrace{\begin{bmatrix} 0 & \frac{I_{L,k} + I_{L,k-1}}{2} & 0 & 0 \\ 0 & 0 & 0 & 0 \\ 0 & 0 & 0 & 0 \\ 0 & 0 & 0 & 0 \end{bmatrix} x_{k-1}}_{f(x_k, I_{L,k}, I_{L,k-1})} - \sum_{j=1}^{k} (-1)^j \gamma_j x_{k+1-j} + w \\ y_k = \underbrace{\begin{bmatrix} -1 & 0 & 1 & -I_{L,k} \end{bmatrix} x_k}_{g(x_k, I_{L,k})} + v \\ \text{where, } \begin{cases} k \le 64, L = k \\ k > 64, L = 64 \end{cases} \end{cases} \tag{15}$$

2.2.2. Model Parameter Identification

In [30], Sierociuk *et al.* put forward the fractional Kalman filter algorithm. In this method, fractional element state estimation, parameter identification and order estimation can be easily realized. Fractional model parameter can be estimated jointly with the state estimation. In order to distinguish the different applications, this method is called fractional joint Kalman filter (FJKF) in this paper. Based on this method, model parameter X_W is regarded as an implicit parameter of the state U_W, the value of X_W can be updated based on the state estimation result. Besides, the predication-correction feature of Kalman filter is effectively used. The application of this algorithm used in battery simplified EIS model parameter identification is described in Algorithm 1.

Algorithm 1 Model parameter identification based on FJKF.

Definitions:

$$A_{k-1} = \left.\frac{\partial f(x_{k-1}, I_{L,k}, I_{L,k-1})}{\partial x_{k-1}}\right|_{x_{k-1}=\hat{x}_{k-1}^{+}} = \begin{bmatrix} 0 & \frac{I_{L,k}+I_{L,k-1}}{2} & 0 & 0 \\ 0 & 0 & 0 & 0 \\ 0 & 0 & 0 & 0 \\ 0 & 0 & 0 & 0 \end{bmatrix}$$

$$C_k = \left.\frac{\partial g(x_k, I_{L,k})}{\partial x_k}\right|_{x_k=\hat{x}_k^{-}} = \begin{bmatrix} -1 & 0 & 1 & -I_{L,k} \end{bmatrix}$$

Step 1: Initialization, Q_k is the covariance of ω_k, R_k is the covariance of noise υ_k. Pk is the error covariance of the state and parameter estimated values, the initial value for each parameter is give as:

$$\hat{x}_0 = E[x],\ P_0^+ = E\left[(x-\hat{x}_0)(x-\hat{x}_0)^T\right]$$

$$R_0 = E\left[v_0 v_0^T\right],\ Q_0 = E\left[w_0 w_0^T\right]$$

Step 2: Time update:

State and parameter time update: $\hat{x}_k^- = f\left(\hat{x}_{k-1}^+, I_{L,k}, I_{L,k-1}\right)$

Error covariance time update: $P_k^- = \left(A_{k-1}+\gamma_1\right)P_{k-1}^+\left(A_{k-1}+\gamma_1\right)^T + Q_x + \sum_{j=2}^{L}\gamma_j P_{k-j}^+ \gamma_j^T$

Step 3: Measurement update:

Kalman gain matrix update: $L_k = P_k^-(C_k)^T\left[C_k P_k^-(C_k)^T + R_k\right]^{-1}$

State and parameter measurement update: $\hat{x}_k^+ = \hat{x}_k^- + L_k^x\left[y_k - g\left(\hat{x}_k^-, I_{L,k}\right)\right]$

Error covariance measurement update: $P_k^+ = \left(I - L_k C_k\right) P_k^-$

Step 4: $k = k + 1$, repeat Step 2 and Step 3, until all data is processed.

2.3. Parameter Identification Experiment Sequence

The battery model and model parameter identification method are described in Section 2.1 andSection 2.2. In order to obtain the battery model parameters and charge capacity, an experimental sequence is designed. The sequence is set up mainly based on dynamic stress tests, and there are seven steps.

The time consumption of this experiment is about eight hours, which is a little longer than the time consumption of constant current/voltage charge and discharge recommended by battery manufacturers, but shorter than the hybrid pulse power test described in the FreedomCAR battery test manual [31]. This sequence is described in Algorithm 2 and Figure 4.

Algorithm 2 Experimental sequence for battery parameter identification.

Step 1: Constant current discharge, until battery terminal voltage reaches the lower cut-off voltage.
Step 2: Rest for one hour.
Step 3: Charge with the battery manufacturer's recommended procedures.
Step 4: Rest for one hour.
Step 5: Apply Dynamic Stress Test on the battery. The maximum discharge current is 2C in this paper. Stop when the battery terminal voltage reaches the lower cut-off voltage. The details of DST are described in FreedomCAR battery test manual.
Step 6: Rest for several minutes.
Step 7: Discharge with a low constant current, until the battery terminal voltage reaches the lower cut-off voltage. This step causes the battery to be fully discharged, this is essential for battery grouping.

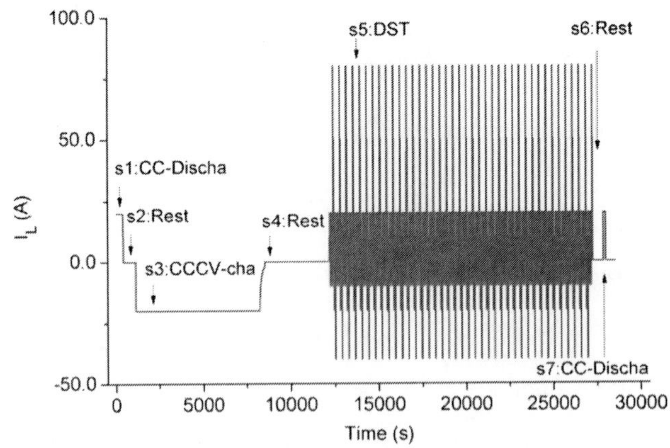

Figure 4. Battery parameter identification experimental sequence.

3. BATTERY FEATURE CLUSTERING

3.1. Data Preprocess

3.1.1. Data Down-Sampling

Battery model parameters U_{OC}, R_O, X_W can be obtained based on the method proposed inSection 2, and they are functions of test time. It is necessary to translate the function into the depth of discharge (*DoD*) of the battery. *DoD* of a battery *i* at time *k* is defined by Equation (16):

$$DoD_i(k) = \frac{\int_{t=0}^{t=k} I_L dt}{Q_i} \times 100\% \tag{16}$$

Here, Q_i is the charge capacity of battery *i* , its value can be calculated by the Ah-counting method, and this process is expressed by Equation (17):

$$Q_i = \int_{step5}^{step7} I_L dt \tag{17}$$

In order to reduce the amount of calculations, a down-sampling process is executed on the parameter identification results. Additionally, model parameter identification results of *DoD* = 0%–10% and 90%–100% are discarded from further analysis.

3.1.2. Crude Data Exclusion

Typically, operational errors are unavoidable in the experimental process. For example, test cables may connect unreliably, equipment channels fail and so on. Apart from these errors, the failures in the battery production process can also lead to the existance of crude values in the model parameter identification results. In this paper, the crude values are discarded based on the Laiyite criterion (3σ criterion). Additionally, since the number of samples remaining should not larger than the designed numbers n_0 for battery grouping, another step is added to the crude data discarding process. The details of this process are shown in Algorithm 3.

Algorithm 3 Crude data excluding process.

Step 1: Calculate the average curve of each model parameters of the samples, additionally, calculate the average value of battery capacities.

$$X_{avg}(DoD) = \frac{1}{n}\sum_{i=1}^{n} X(DoD)$$

Where, X represents Uoc, Ro or Xw.

Step 2: Calculate the average distance between the parameter curve and the average curve, that is the distance between battery capacities and the average capacity.

$$D_{X,i} = mean\left(X_i(DoD) - X_{avg}(DoD)\right), i = 1,...,n$$

Step 3: Calculate the standard deviation of each distance parameter.

$$\sigma_X = \sqrt{\frac{1}{n}\sum_{i=1}^{n} D_{X,i}^{\ 2}}$$

Step 4: Discard the samples that do not meet the requirement of Laiyite criterion. The Laiyite criterion is: $|DX_i| < 3\sigma_X$

Step 5: Regard the remaining samples as the evolution objects. Reset the number of evolution objects n, and repeat Steps 1–4, until all the remaining parameters are eligible with the requirement of Laiyite criterion.

Step 6: If $n > n_0$, randomly discard the samples which have maximum value of $|DX_i|$ until the number of remaining samples is equal to n_0. Ultimately, the samples used for battery grouping are obtained.

3.2. Consistency Evaluation Parameter Generation

In Section 3.1, D_Q, $D_{U_{OC}}$, D_{R_O} and D_{X_W} of each remaining battery are obtaine; these parameters have different units and should be normalized for further analysis, the results are represented by $D^*_{U_{OC}}$, $D^*_{R_O}$ and $D^*_{X_W}$. The four parameters describe the performance of batteries, but cannot be used directly for battery consistency evaluation, because they have different effects on battery consistency. It is necessary to weight the parameters according to defined rules. In this paper, the weights are set up based on the parameters' inconsistent influence on the terminal voltage. Battery terminal voltage can be expressed as:

$$U_T = U_{OC} - U_{R_O} - U_W \tag{18}$$

The relationship between Q's inconsistent value ΔQ and ΔU_{OC} is described in Figure 5. In order to simplify the calculation, the OCV curve can be regard as a straight line, and Equation (19) can be obtained.

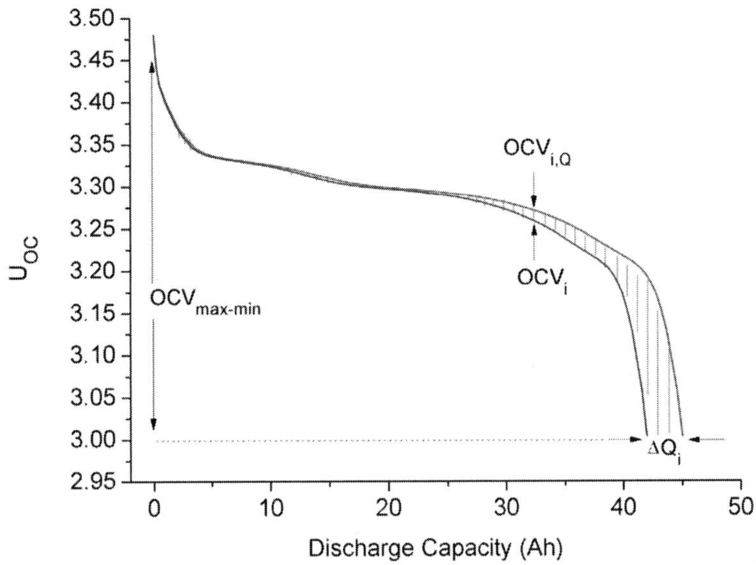

Figure 5. Effect of capacity inconsistency on U_{OC} (U_T).

$$\Delta U_{T,Q} = \frac{\frac{1}{2}\Delta OCV_{\max-\min}\Delta Q}{Q_{avg}} \tag{19}$$

where, $\Delta OCV_{\max-\min}$ is about 0.48 V.

The relationship between ΔU_T and ΔU_{OC}, ΔR_O can be easily obtained; this are expressed as:

$$\Delta U_{T,U_{OC}} = \Delta U_{OC} \tag{20}$$

$$\Delta U_{T,R_O} = I_L \Delta R_O \tag{21}$$

However, the influence of ΔX_W on ΔU_T cannot be obtained directly. Thus, the following analysis is carried out. Assuming the initial state of U_W is zero, when a step current excitation I_L is applied to Z_W, the voltage response after k s can be calculated. When k = 1, the voltage response is given as:

$$U_{W,1} = X_W I_L - (-1)^1 \binom{0.5}{1} U_{W,0} \tag{22}$$

The voltage response of Z_W when $k \geq 2$ is given as:

$$U_{W,k+1} = X_W I_L - \sum_{j=1}^{L} (-1)^j \binom{0.5}{j} U_{W,k+1-j} \tag{23}$$

Finally, the voltage response of Z_W at time k can be given as:

$$U_{W,T} = a(k) X_W I_L \tag{24}$$

where, $a(k)$ is coefficient variable of $X_W I_L$, and is determined by L in the state discrete equations. I_L and X_W are constant, the coefficient a can be expressed in Figure 6. The maximum value of a is $a_{max}(k) = 14$.

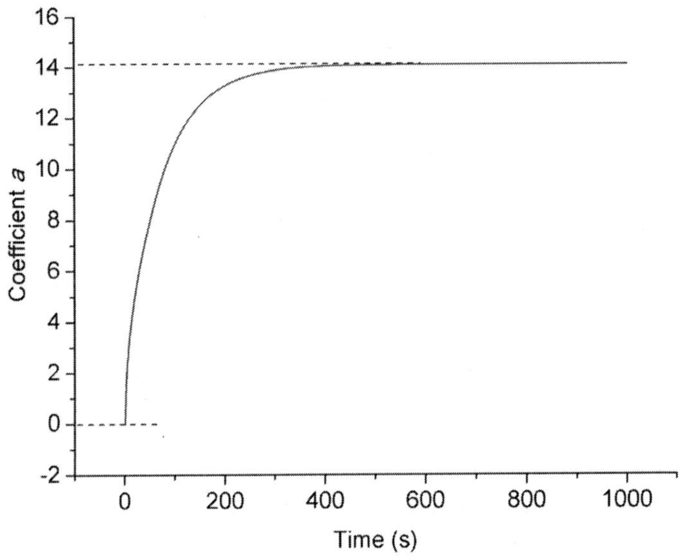

Figure 6. Changes of the coefficient *a* with time.

During charge or discharge, the working time is much longer than the length of time presented in the figure. $a_{max} = 14$, the value of a changes mainly near the maximum value. The relationship between ΔU_T and ΔX_W can be described as:

$$\Delta U_W = a_{max} I_L \Delta X_W \tag{25}$$

The median working current 1C (40 A) is used as the reference current $I_{L,ref}$ in the Equations (24) and (25), the weight proportion of each parameter can be described as:

$$w_Q : w_{U_{OC}} : w_{R_O} : w_{X_W} = \frac{\frac{1}{2} OCV_{\max-\min} \Delta Q}{Q_0} : \Delta U_{OC} : I_{L,ref} \Delta R_O : a_{\max} I_{L,ref} \Delta X_W \tag{26}$$

After the parameters are weighted, the feature vector Cell $\left(D^*_{Q,w}, D^*_{U_{OC},w}, D^*_{R_O,w}, D^*_{X_W,w}\right)$ which consists of four consistency evaluation parameters for each cell is obtained.

3.3. Battery Equal-Number Clustering

Battery grouping is a special kind of clustering problem. In the traditional clustering method, elements are grouped based on the similarity of feature data, and there is no pre-set category features. The whole clustering process works without supervision. However, battery grouping is a little different from the traditional method, because the amount of elements in each group should meet the requirements of the battery pack. Hence, a special equal-number clustering method is needed for battery grouping.

3.3.1. K-Means Clustering Method

K-means clustering method is a popular partitioning clustering method [32]. In this method, the initial cluster centers are randomly selected from the elements to be clustered, and the remaining elements are clustered into *N* groups based on the similarity (distance), and the cluster center of each cluster is recalculated after that. The variance function shown in Equation (27) is generally used as the standard measurement function in this method. The main steps of K-means are shown in Algorithm 4.

$$E = \sum_{n=1}^{N} \sum_{x \in c_n} \left\| x - \mu_n \right\|^2 \tag{27}$$

Algorithm 4 Main steps of K-means clustering method.

Step 1: Initialize the cluster centers. Select N cluster centers randomly. $\mu_1, \mu_2, \ldots, \mu_k$.

Step 2: Calculate the distance between each sample and each cluster center. Clustering the samples to the nearest cluster based on the minimum distance principle. For sample *p*, it is belong to the cluster which is obtained by: $c^{(p)} = \arg\min \left\| x^{(p)} - \mu_n \right\|^2$

Step 3: Recalculate the cluster centers. The mean value of each cluster is used as the new cluster center:

$$\mu_n = \frac{\sum_{p=1}^{M} 1\left\{c^{(p)} = n\right\} x^{(p)}}{\sum_{p=1}^{M} 1\left\{c^{(p)} = n\right\}}$$

Step 4: Repeat Step 2 and Step 3, until the cluster centers are no longer changing or only changes minimal.

3.3.2. Kd-Tree Cluster Center Initialization

Unfortunately, K-means clustering method is local optimization strategy, and it is sensitive to the initial cluster center and can easily fall into a local optimal solution. A good initial cluster center can avoid the local optima issue. Redmond, *et al.* proposed a cluster center initial method based on kd-tree. kd-tree is established for data density estimation, and the initial cluster centers are selected from kd-tree's leave boxes based on an improved max-min distance method. This cluster center initial method is much better than the random cluster center initial method. The calculation is relatively simple. Readers can refer to [33] for further details of the algorithm.

3.3.3. The New Modified K-Means Clustering Method

Section 3.3.1 and Section 3.3.2 describe the K-means clustering method using kd-tree cluster center initialization. The battery equal-number clustering method proposed in this paper is based on this method and the clustering rules are redesigned. The target of the method is partitioning the samples into N clusters and the amount of elements in each cluster is equal to M. The objective function of equal-number clustering is the same with the traditional K-means method, as shown in Equation (27). An adjusting rule is developed to equalize the numbers as below:

Rule 1: Element number equalization

The cluster which has the largest number of elements should be determined first. Then, the elements of this cluster are sorted based on the distances between them and their cluster center. The element with the farthest distance is assigned to another cluster whose number of elements is less than the pre-set value M and the distance of this element to the new cluster should be as near as possible. This assigning rule takes distance and element number into account at the same time. We repeat this process until the number of elements in each cluster is equal to the pre-set value. Equal-number clusters can be accomplished based on Rule 1. However, this process cannot guarantee the accuracy of the element's attribution. Thus, an element exchange criterion between two clusters is set up.

Rule 2: Element exchange criterion

This rule is made to regulate the element exchanging process between two selected clusters. In order to explain this rule, several definitions are made:

Master cluster Cm: Cm is the cluster who initiates the element exchange process. The cluster center and element i in this cluster is expressed by μ_m and $x_{m,i}$.

Slave cluster Cs: Cs is the cluster waiting for element exchange with Cm. The cluster center and element j is expressed μs and xs,j.

Master (Slave) center vector $\overrightarrow{\mu_m\mu_s}$ ($\overrightarrow{\mu_s\mu_m}$): The vector's starting point is μ_m (μ_s) and ending point is μ_s(μ_m).

Master (Slave) proximal element $x_{m,ns}$ ($x_{s,nm}$): The element of C_m (C_s) near μ_s (μ_m). If $\overrightarrow{\mu_m x_{m,i}} \cdot \overrightarrow{\mu_m\mu_s} > 0$, element i is a Master proximal element Similarly, If $\overrightarrow{\mu_s x_{s,j}} \cdot \overrightarrow{\mu_s\mu_m} > 0$, element *j* is a slave proximal element.

In these two clusters, the exchange process is only allowed in the proximal elements because the exchange possibility between two proximal elements is much higher than between non-proximal elements. This limitation is an effective approach to reduce calculation complexity and improve the convergence rate of the clustering algorithm.

In order to decrease the value of the measurement target of the new modified K-means method, exchange process between C_m and C_s occurs when the requirements of Equation (28) is satisfied:

$$\left\|x_{m,i}-\mu_m\right\|^2+\left\|x_{s,j}-\mu_s\right\|^2>\left\|x_{s,j}-\mu_m\right\|^2+\left\|x_{m,i}-\mu_s\right\|^2 \quad (28)$$

Based on the analysis and two new rules for cluster elements attribution, the new modified K-means method is established. This equal-number clustering steps are described in the Algorithm 5 and some of the main steps are shown in Figure 7.

Algorithm 5 Steps of equal-number clustering method.

Step 1: Cluster centers initialization. Cluster center is calculated using the kd-tree cluster center initialization method.

Step 2: Cluster initialization. The elements are clustered to each cluster according to the minimum distance criterion described in K-means clustering method. Step1 and Step 2 are shown in Figure 7a.

Step 3: Element number equalization. Equalize the number of each cluster according to rule 1. This step is shown in Figure 7b.

Step 4: Distance data storage. Two arrays are set up for each cluster, named main storage array and auxiliary storage array. Calculate the distances from each element to all centers. The results are stored in the main storage arrays, but auxiliary storage array is not used here.

Step 5: Elements exchange among the clusters.

Step 5.1: Set one of the clusters as a master cluster C_m, another one is regard as a slave cluster C_s. Find the proximal elements in these two clusters. This step is shown in Figure 7c.
Step 5.2: Select a proximal element in C_m and a proximal element in C_s as the quasi-exchange elements, If the two elements can meet the requirements of Equation (28), take out the two elements from their pervious clusters, exchange the belonging of elements and store their feature parameters in the auxiliary storage array. If not, select another proximal element in C_s as the quasi-exchange element and repeat the calculation of Equation (28) and determine whether to exchange elements or not, until all $x_{s,nm}$s are traversed. This step is shown in Figure 7d.
Step 5.3: Select another proximal element in C_m as the quasi-exchange element, repeat Step 5.2, until all $x_{m,ns}$ s are traversed.
Step 5.4: Set another cluster as a new C_s, repeat Step 5.2 and Step 5.3, until all clusters apart from the master cluster are treated as a slave cluster one time for element exchange.
Step 5.5: Set another cluster as a new C_m, repeat Step 5.2, Step 5.3 and Step 5.4, until all the clusters are treated as a master cluster one time.

Step 6: Take out the data stored in the auxiliary storage array, and push them onto the main storage array of each cluster. Step 6: Recalculate the center of each cluster. This step is shown in Figure 7e.

$$\mu_n = \frac{\sum_{p=1}^{M} 1\left\{c^{(p)} = n\right\} x^{(p)}}{\sum_{p=1}^{M} 1\left\{c^{(p)} = n\right\}}$$

Step 7: Repeat Steps 4–6, until cluster centers do not change or only change a little.
$\sum \|\Delta\mu_n\| < \varepsilon$
Where, ε is the threshold value of process stopping, the equal-number clusters are obtained.

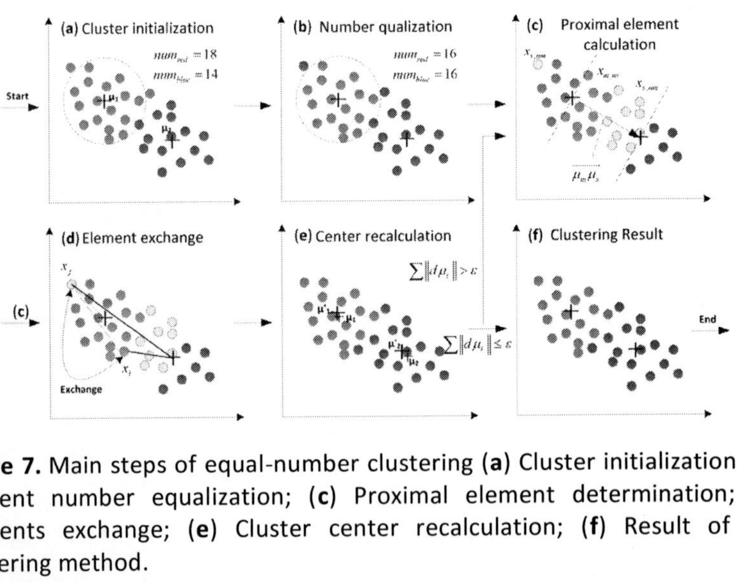

Figure 7. Main steps of equal-number clustering (**a**) Cluster initialization; (**b**) Element number equalization; (**c**) Proximal element determination; (**d**) Elements exchange; (**e**) Cluster center recalculation; (**f**) Result of the clustering method.

4. EXPERIMENTAL DETAILS

Battery parameter identification and battery grouping verification experiments were implemented at CALB Co. Ltd. (Luoyang, China). A series of battery test systems were used for battery testing. Model and parameter identification verification experiments were carried out with an Arbin BT2000 battery test system in the Harbin Institute of Technology. The devices under test are 95 energy-type $LiFePO_4$/graphite batteries produced by CALB. The typical capacity (Q_0) is 42 Ah, cut-off voltages are 3.65 V and 2.5 V, maximum constant discharge current is 2C. The experiments were conducted at room temperature and standard atmospheric pressure.

4.1. Parameter Identification Experimental Details

The sequence of parameter identification experiment has been described in Section 2.3, this sequence is applied to the 95 batteries. The terminal voltage and load current value are recorded in the experiment.

4.2. Verification Experimental Details

There are two verification experiments in this paper. One is a model and parameter identification validation experiment, the other is a battery grouping verification experiment.

4.2.1. Model and Parameter Identification Verification Experiments

In order to validate the accuracy of the simplified EIS model and parameter identification method, a load current sequence combining the DST profile with constant current discharge profile is established. The change of *DoD* in each segment is 10%, the interval of every segment is 1 h. This sequence is shown in Figure 8.

A1: The segment of the test data in *DoD* = 50% is selected for model accuracy verification. U_{OC}only changes slightly in this segment, and the value of U_{OC} can be regarded as constant. The simulation results of the first-order RC model and the simplified EIS model are compared. Genetic algorithm (GA) is used for model parameter identification in this process.

A2: The terminal voltage of each segment's first point is regarded as the value of U_{OC} at the initial time. The value of U_{OC} at the end point of this segment is determined by the terminal voltage of the next segment's first point. The change of U_{OC} is considered to be linear with the change of *DoD* in the center of this segment. GA and FJKF are used to get the impedance parameters in the model. The accuracy of FJKF is verified by comparing the parameter identification results of the two methods.

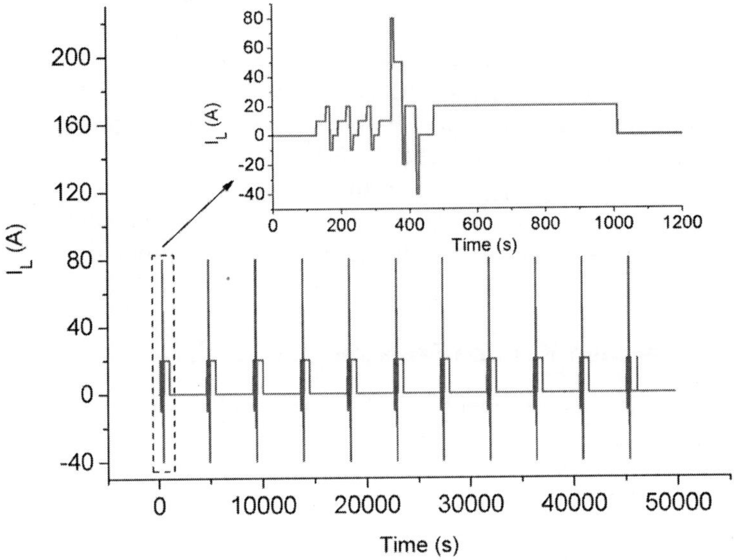

Figure 8. Model and parameter identification verification experimental sequence.

4.2.2. Battery Grouping Verification Experiments

In order to verify the accuracy of the battery grouping method, typically, batteries should be connected in parallel or series and experiments performed. In this paper, the batteries are not connected in parallel or in series because the analysis is too complex and meaningful results may not be obtained. Instead of that, five simple experiments are implemented on each battery respectively. Battery grouping accuracy is evaluated by the aggregation degree of the voltage curves in each group. The experiments include: 0.5 C-rate charge and 0.5 C-rate discharge, 1 C-rate charge and 1 C-rate discharge and dynamic stress test.

5. RESULTS AND DISCUSSION

5.1. Battery Grouping Result Analysis

5.1.1. Battery Parameter Identification Result Analysis

Battery model parameters are extracted from the experiment using the FJKF algorithm described in Section 2.3, and battery capacity is obtained using Ah-counting. Battery capacity and its distribution are shown in Figure 9, where the capacity values are located between 40 Ah and 44 Ah, and the distribution is approximately normal. Battery model parameters and terminal voltage estimation errors are shown in Figure 10.

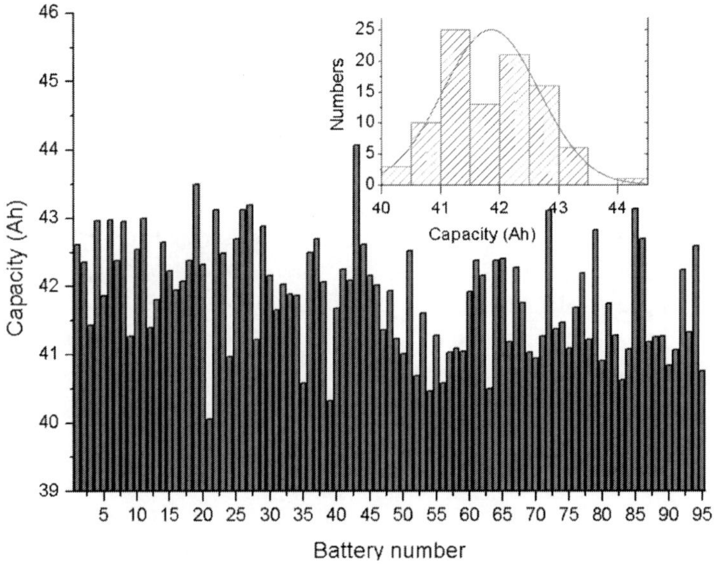

Figure 9. Battery capacities and distribution.

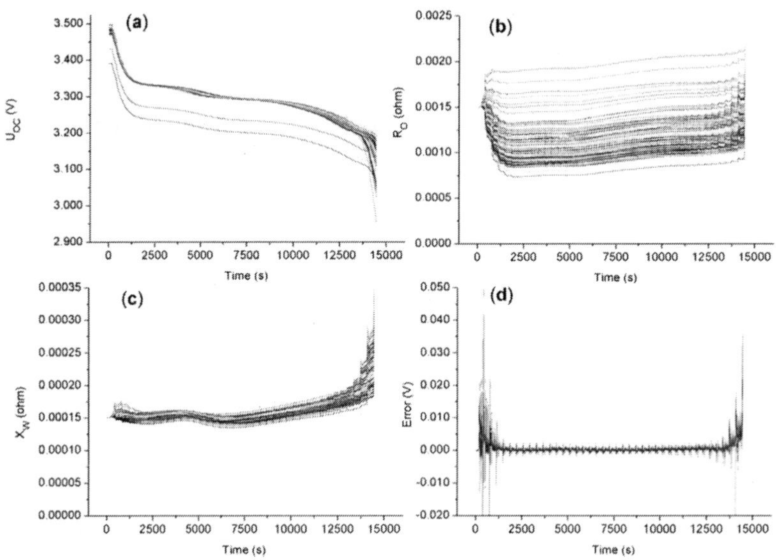

Figure 10. Model parameter identification result (**a**) U_{OC}; (**b**)R_O; (**c**) X_W; (**d**) U_T error.

In Figure 10a,b, there exist apparent crude values in the parameter curves. In Figure 10d, the terminal voltage estimation errors in most regions (*DoD* = 10%–90%) are smaller than 5 mV. This phenomenon can partly verify the accuracy of the simplified EIS model and FJKF algorithm. The parameters in the regions of *DoD* = 0%–10% and *DoD* = 90%–100% is unreliable because of the large simulation error existing in these regions. The data are discarded.

5.1.2. Battery Clustering Result Analysis

Model parameters are down sampled and the crude data is discarded according to the data preprocessing process described in Section 3.1. The result is shown in Figure 11. Then, the weights of the battery performance parameters are calculated based on Equation (26). The values of$OCV_{max-min}$, $I_{L,ref}$ and a_{max} are obtained in Section 3.2 and the other coefficients in this equation can be obtained from the preprocess result shown in Figure 11:

$$\begin{cases} \Delta Q = 3.5 \text{ Ah} \\ \Delta U_{OC} = 8.0 \times 10^{-3} \text{ V} \\ \Delta R_O = 5.0 \times 10^{-4} \ \Omega \\ \Delta X_W = 1.2 \times 10^{-5} \ \Omega \end{cases} \quad (29)$$

Then, the weights can be obtained. After normalizing the minimum weight and rounding the value of each weight, Equation (30) can be obtained:

$$w_Q : w_{U_{OC}} : w_{R_O} : w_{X_W} \approx 3:1:3:1 \quad (30)$$

Battery clustering result are shown in Figure 12. The samples are clustered into four groups. In each group, the consistency evolution parameters are very similar with each other.

5.2. Verification of the Battery Grouping Method

5.2.1. Model Accuracy Verification

In this section, the first order RC model is compared with the simplified EIS model. In order to obtain the model parameters, a genetic algorithm (GA) is used to find the optimal value of the parameters except for U_{OC}. Equation (31) is used as the objective function of GA.

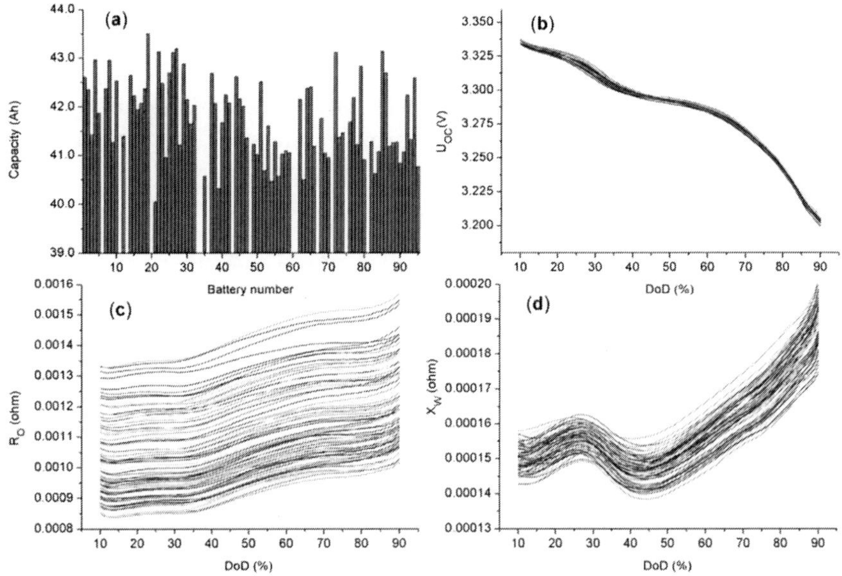

Figure 11. The characteristic parameters after preprocess (**a**) Q; (**b**) U_{OC}; (**c**) R_O; (**d**) X_W.

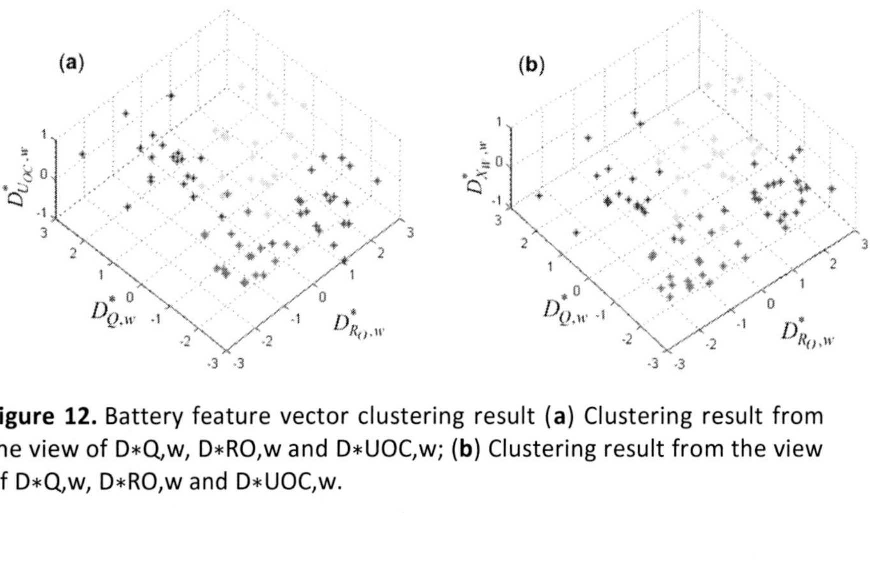

Figure 12. Battery feature vector clustering result (**a**) Clustering result from the view of D∗Q,w, D∗RO,w and D∗UOC,w; (**b**) Clustering result from the view of D∗Q,w, D∗RO,w and D∗UOC,w.

$$f(X) = \frac{1}{N}\sum_{k=1}^{N}\left(U_{T,k} - \widehat{U_{T,k}}\right)^2 \tag{31}$$

The measured terminal voltages of DST in *DoD* = 50% are used as the true values. The optimization results of model parameters are used for terminal voltage estimation and predication. Results are shown in Figure 13.

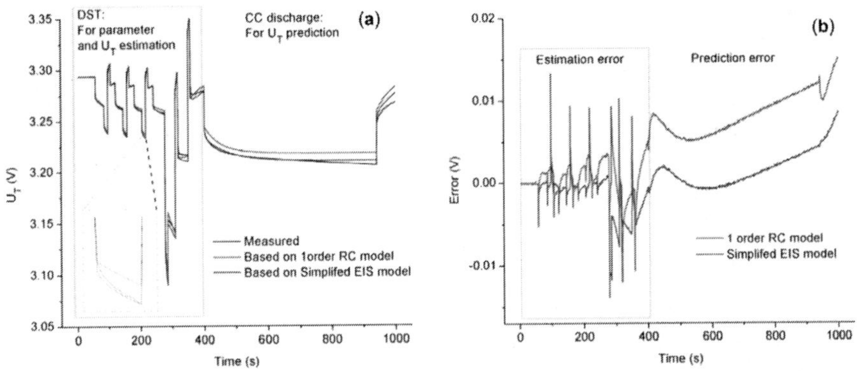

Figure 13. Comparison of battery model simulation results (**a**) simulation results and measured values of terminal voltage; (**b**) simulation errors.

In the test results, the root mean square (RMS) voltage error of simulation results of the first-order RC model and the simplified EIS model are 6.9 mV and 2.8 mV. The estimation errors in DST region are 2.7 mV and 2.8 mV, and prediction errors are 8.7 mV and 2.7 mV. The simulation accuracy of the simplified EIS model is higher than the accuracy of the first-order RC model. Compared to the first-order RC model, the simplified EIS model can represent the impedance characteristics of the battery much better. Furthermore, the number of parameters is less than that of the first-order RC model.

5.2.2. Model Parameter Identification Accuracy Verification

The RMS error of terminal voltage is 1.03 mV in the result which is shown in Figure 9. In this section, FJKF and GA are applied to the same experimental data for model parameter identification. The identification results of U_{OC}, R_O and X_W are shown in Figure 14. The results of the two methods are very similar. The accuracy and effectiveness of FJKF can be verified from the results.

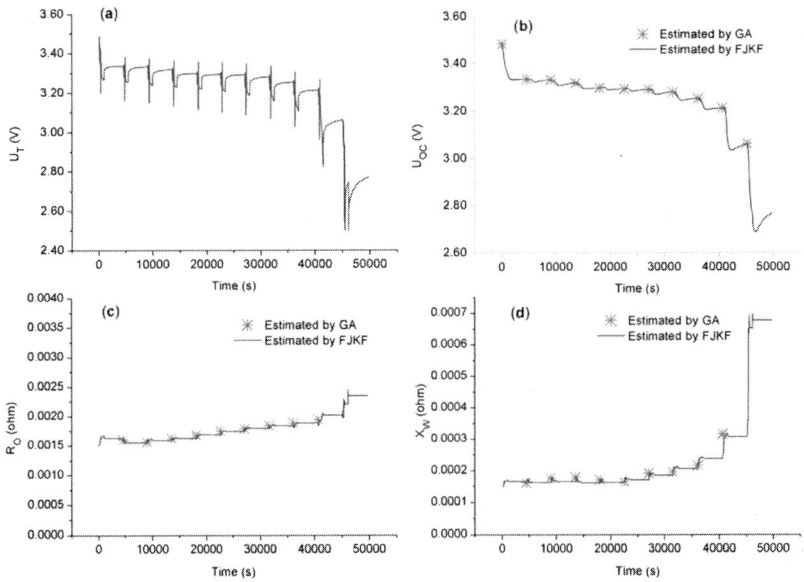

Figure 14. Comparison of the model parameter identification methods: (**a**) terminal voltage; (**b**) U_{OC}; (**c**) R_O; (**d**) X_W.

5.2.3. Battery Grouping Accuracy Verification

In order to verify the validity of the proposed method for battery grouping, three other battery grouping methods—capacity matching, R_O matching and the voltage curve matching method—are used to compare with this method. The first two methods are easy to implement. The voltage curve matching method, proposed in [34], is based on the similarity of average distance, Euclidean distance and correlation coefficient, but the clustering method for three parameters was not described. In this paper, the voltage curve matching method is based on the average distance between the voltage curves and the mean curve. Additionally, 1C-rate discharge curves are used for battery grouping in this method. The clustering effective evaluation criterion is dominated based on the average Euclidean distance of the battery voltage curves of each cluster. The evaluation equation is shown as:

$$E\left(\text{group}_i\right)=\frac{1}{N}\sum_{n=1}^{N}\sqrt{\frac{1}{K}\sum_{k=1}^{K}\left(U_{T,i}\left(k\right)-U_{T,g_n}^{avg}\left(k\right)\right)^2} \tag{32}$$

The battery grouping evaluation results of these methods are shown in Table 1. As is seen from the results, the voltage dispersion in each group after adopting the proposed clustering method is much smaller than the other ones under most working conditions, so the new grouping method has relatively higher accuracy when compared with the other methods.

Table 1. Comparison of battery grouping results.

Method	Capacity Matching ($\times 10^{-3}$ V)			Ro Matching ($\times 10^{-3}$ V)			Curve Matching ($\times 10^{-3}$ V)			Proposed Method ($\times 10^{-3}$ V)		
Dispersion	Max	Min	Avg.	Max	Min	Avg.	Max	Min	Avg.	Max	Min	Avg.
DST	21.56	9.483	15.40	38.43	19.18	25.94	16.50	6.910	13.41	**15.80**	**5.682**	**10.78**
0.5 C charge	26.02	9.406	17.00	43.61	27.31	32.71	22.78	7.803	18.22	**19.60**	**6.425**	**13.98**
0.5 C discharge	21.53	9.391	15.31	41.90	26.13	31.23	19.40	7.776	15.82	**19.17**	**7.548**	**14.21**
1 C charge	30.35	15.88	24.40	42.65	28.98	33.67	**22.74**	17.89	20.32	22.93	**16.85**	**19.64**
1 C discharge	24.40	16.13	20.46	42.92	27.97	32.32	23.32	**9.052**	17.03	**21.80**	9.939	**16.71**

6. CONCLUSIONS

A grouping method for $LiFePO_4$/graphite batteries is proposed. The method is mainly based on the fractional joint Kalman filter and a new modified K-means algorithm. Batteries can be divided into groups with same number of elements and similar battery characteristics in each group. The innovations and contributions of the proposed method are as follows:

1. A simplified EIS battery model is adopted for battery characteristic description. The accuracy of this model is relatively higher and the number of parameters is less than in the first order RC model under the tested conditions. The model is suitable for battery grouping.

2. A fractional joint Kalman filter is used for parameter identification. Fractional parameters $X_W$can be estimated jointly with the state U_W of the fractional element Z_W. This algorithm is appropriate and effective for fractional model parameter identification.

3. A new modified K-means clustering method is proposed for battery equal-number grouping. In this method, a feature vector with four parameters is used to describe the performance of the battery from different angles. Two rules are designed to equalize the numbers of elements in each group and exchange samples among groups. The rules are added to the K-means clustering method and replace the criterion of element attribution calculation.

This battery grouping method is effective for $LiFePO_4$/graphite battery grouping. Furthermore, this method has the potential to be utilized in battery reuse applications. The simplified EIS model and FJKF algorithm can also be used for battery online state estimation or prediction. The equal-number clustering method can be applied to other clustering applications which require not only similarity, but also equal-numbers in each cluster. Future works will include evaluating the algorithm for the abovementioned applications.

ACKNOWLEDGMENTS

This research was supported by the Research and Development of Application Technology Plan Project in Heilongjiang Province of China (GA13A202), the NSFC-EPSRC Collaborative Research Initiative in Smart Grids and the Integration of Electric Vehicles (51361130153), and the Science and Technology Project of State Grid Corporation of China. The authors thank the support of China Aviation Lithium Battery (CALB) Co., LTD, Teng Wang and Yanyan Zhao's help on this research. The author would also like to acknowledge the reviewers' corrections and helpful suggestions.

AUTHOR CONTRIBUTIONS

Xiaoyu Li designed the algorithm, some parts of the experiment and wrote the main parts of the manuscript. Kai Song checked the whole manuscript and gave some suggestions. Guo Wei and Rengui Lu designed some parts of experiments. Chunbo Zhu checked the results and the whole manuscript.

REFERENCES

1. Wang, T.; Tseng, K.J.; Zhao, J.; Wei, Z. Thermal investigation of lithium-ion battery module with different cell arrangement structures and forced air-cooling strategies. *Appl. Energy***2014**, *134*, 229–238.

2. Xiong, B.; Zhao, J.; Wei, Z.; Skyllas-Kazacos, M. Extended Kalman filter method for state of charge estimation of vanadium redox flow battery using thermal-dependent electrical model. *J. Power Sources* **2014**, *262*, 50–61.

3. Wang, T.; Zhu, C.; Pei, L.; Lu, R.; Xu, B. The State of Arts and Development Trend of SOH Estimation for Lithium-Ion Batteries. In Proceedings of the 2013 IEEE Vehicle Power and Propulsion Conference (VPPC), Beijing, China, 15–18 October 2013; pp. 1–6.

4. Andre, D.; Appel, C.; Soczka-Guth, T.; Sauer, D.U. Advanced mathematical methods of SOC and SOH estimation for lithium-ion batteries. *J. Power Sources* **2013**, *224*, 20–27.

5. Pei, L.; Zhu, C.; Wang, T.; Lu, R.; Chan, C.C. Online peak power prediction based on a parameter and state estimator for lithium-ion batteries in electric vehicles. *Energy* **2014**, *66*, 766–778.

6. Xiong, R.; He, H.; Sun, F.; Liu, X.; Liu, Z. Model-based state of charge and peak power capability joint estimation of lithium-ion battery in plug-in hybrid electric vehicles. *J. Power Sources* **2013**, *229*, 159–169.

7. Kim, J.; Shin, J.; Chun, C.; Cho, B.H. Stable Configuration of a Li-Ion Series Battery Pack Based on a Screening Process for Improved Voltage/SOC Balancing. *IEEE Trans. Power Electron.* **2012**, *27*, 411–424.

8. Kim, J.; Cho, B.H. Screening process-based modeling of the multi-cell battery string in series and parallel connections for high accuracy state-of-charge estimation. *Energy* **2013**, *57*, 581–599.

9. Schneider, E.L.; Oliveira, C.T.; Brito, R.M.; Malfatti, C.F. Classification of discarded NiMH and Li-Ion batteries and reuse of the cells still in operational conditions in prototypes. *J. Power Sources* **2014**, *262*, 1–9.

10. Fang, K.; Chen, S.; Mu, D.; Wu, B.; Wu, F. Investigation of nickel–metal hydride battery sorting based on charging thermal behavior. *J. Power Sources* **2013**, *224*, 120–124.

11. Li, X.; Wang, T.; Pei, L.; Zhu, C.; Xu, B. A comparative study of sorting methods for lithium-ion batteries. In Proceedings of the 2014 IEEE Conference and Expo Transportation Electrification Asia-Pacific (ITEC Asia-Pacific), Beijing, China, 31 August–3 September 2014; pp. 1–6.

12. Diamond, R.A.; Wang, H.; Chen, F.; Wilke-Douglas, M. Cell Preparation and Enrichment for FCM Analysis and Cell Sorting. In *In Living Color*; Springer Lab Manuals; Springer: Berlin/Heidelberg, Germany, 2000; pp. 111–141.

13. Guo, L.; Liu, G.W. Research of Lithium-Ion Battery Sorting Method Based on Fuzzy C-Means Algorithm. *Adv. Mater. Res.* **2012**, *354–355*, 983–988.

14. Yun, W.U.; Ge-Chen, L.I.; Chen, S.L. The Application of BP Nerual Network in Battery-sorting. *J. Harbin Univ. Sci. Technol.* **2001**, *5*, 54–57.

15. Feng, F.; Lu, R.; Zhu, C. A Combined State of Charge Estimation Method for Lithium-Ion Batteries Used in a Wide Ambient Temperature Range. *Energies* **2014**, *7*, 3004–3032.

16. Plett, G.L. Extended Kalman filtering for battery management systems of LiPB-based HEV battery packs Part 1. Background. *J. Power Sources* **2004**, *134*, 262–276.

17. Plett, G.L. Extended Kalman filtering for battery management systems of LiPB-based HEV battery packs Part 2. Modeling and identification. *J. Power Sources* **2004**, *134*, 252–261.

18. Plett, G.L. Extended Kalman filtering for battery management systems of LiPB-based HEV battery packs Part 3. State and parameter estimation. *J. Power Sources* **2004**, *134*, 277–292.

19. Xiong, R.; Sun, F.; He, H.; Nguyen, T.D. A data-driven adaptive state of charge and power capability joint estimator of lithium-ion polymer battery used in electric vehicles. *Energy***2013**, *63*, 295–308.

20. Yoon, S.; Hwang, I.; Lee, C.W.; Ko, H.S.; Han, K.H. Power capability analysis in lithium ion batteries using electrochemical impedance spectroscopy. *J. Electroanal. Chem.* **2011**, *655*, 32–38.

21. Xu, J.; Mi, C.C.; Cao, B.; Cao, J. A new method to estimate the state of charge of lithium-ion batteries based on the battery impedance model. *J. Power Sources* **2013**, *233*, 277–284.

22. Waag, W.; Käbitz, S.; Sauer, D.U. Experimental investigation of the lithium-ion battery impedance characteristic at various conditions and aging states and its influence on the application. *Appl. Energy* **2013**, *102*, 885–897.

23. Luo, W.; Lyu, C.; Wang, L.; Zhang, L. A new extension of physics-based single particle model for higher charge–discharge rates. *J. Power Sources* **2013**, *241*, 295–310.

24. Fuller, T.F.; Doyle, M.; Newman, J. Relaxation phenomena in lithium-ion-insertion cells. *J. Electrochem. Soc.* **1994**, *141*, 982–990.

25. Zhang, D. Modeling Lithium Intercalation of a Single Spinel Particle under Potentiodynamic Control. *J. Electrochem. Soc.* **1999**, *147*, 831–838.

26. Zhang, L.; Lyu, C.; Hinds, G.; Wang, L.; Luo, W.; Zheng, J.; Ma, K. Parameter Sensitivity Analysis of Cylindrical $LiFePO_4$ Battery Performance Using Multi-Physics Modeling. *J. Electrochem. Soc.* **2014**, *161*, A762–A776.

27. Fleischer, C.; Waag, W.; Heyn, H.; Sauer, D.U. On-line adaptive battery impedance parameter and state estimation considering

physical principles in reduced order equivalent circuit battery models part 2. Parameter and state estimation. *J. Power Sources* **2014**, *262*, 457–482.

28. Fleischer, C.; Waag, W.; Heyn, H.; Sauer, D.U. On-line adaptive battery impedance parameter and state estimation considering physical principles in reduced order equivalent circuit battery models : Part 1. Requirements, critical review of methods and modeling. *J. Power Sources* **2014**, *260*, 276–291.

29. Li, X.; Zhu, C.; Wei, G.; Lu, R. Online Parameter Estimation of $LiFePO_4$ Battery Simplified Impedance Spectroscopy Model Based on Fractional Joint Kalman Filter. *Trans. Chin. Electrotechnical Soc.* **2015**. in press.

30. Sierociuk, D.; DzielińSki, A. Fractional Kalman filter algorithm for the states, parameters and order of fractional system estimation. *Int. J. Appl. Math. Comput. Sci.* **2006**, *16*, 129–140.

31. *FreedomCAR Battery Test Manual for Power-Assist Hybrid Electric Vehicles*; U.S. Department of Energy: Washington, DC, USA, 2003.

32. Mcqueen, J. Some methods for classification and analysis of ultivariate observations. In Proceedings of the Berkeley Symposium Mathematical Statistics Probability; University of California Press: Berkeley, CA, USA, 1967; pp. 281–297.

33. Redmond, S.J.; Heneghan, C. A Method for Initialising the K-Means Clustering Algorithm Using Kd-Trees. *Pattern Recognit. Lett.* **2007**, *28*, 965–973.

34. Hanze, D.Z.L.G. Cell classification system based on automatic curve-recognition. *Chin. J. Power Sources* **2000**, *2*, 65–69.

Chapter 6

An Improved PDR/Magnetometer/Floor Map Integration Algorithm for Ubiquitous Positioning Using the Adaptive Unscented Kalman Filter

Jian Wang[1,*], Andong Hu[1,2], Xin Li[3] and Yan Wan[4]

[1]*School of Environmental Science and Spatial Informatics, China University of Mining and Technology, Xuzhou 221116, China*

[2]*School of Mathematical and Geospatial Sciences, Royal Melbourne Institute of Technology University, Melbourne, Vic 3001, Australia*

[3]*School of Computer Science and Technology, China University of Mining and Technology, Xuzhou 221116, China*

[4]*Chemical Engineering Institute, China University of Mining and Technology, Xuzhou 221116, China*

ABSTRACT

In this paper, a scheme is presented for fusing a foot-mounted Inertial Measurement Unit (IMU) and a floor map to provide ubiquitous positioning in a number of settings, such as in a supermarket as a shopping guide, in a fire emergency service for navigation, or with a hospital patient to be tracked. First, several Zero-Velocity Detection (ZDET) algorithms are compared and discussed when used in the static detection of a pedestrian. By introducing information on the Zero Velocity of the pedestrian, fused with a magnetometer measurement, an improved

Pedestrian Dead Reckoning (PDR) model is developed to constrain the accumulating errors associated with the PDR positioning. Second, a Correlation Matching Algorithm based on map projection (CMAP) is presented, and a zone division of a floor map is demonstrated for fusion of the PDR algorithm. Finally, in order to use the dynamic characteristics of a pedestrian's trajectory, the Adaptive Unscented Kalman Filter (A-UKF) is applied to tightly integrate the IMU, magnetometers and floor map for ubiquitous positioning. The results of a field experiment performed on the fourth floor of the School of Environmental Science and Spatial Informatics (SESSI) building on the China University of Mining and Technology (CUMT) campus confirm that the proposed scheme can reliably achieve meter-level positioning.

Keywords: zero-velocity detection; adaptive unscented Kalman filter; heading angle; floor map matching; Inertial Measurement Unit; Pedestrian Dead Reckoning

1. INTRODUCTION

Pedestrian Dead Reckoning (PDR) [1,2] is a priority technique for the first responder ubiquitous positioning system. Several positioning algorithms have been developed over the past ten years.

The basic PDR model involves three components: step detection, step length estimation and heading angle estimation [3,4]. Acceleration measurements are an ideal choice for step detection because of the periodicity of a pedestrian's walking pattern. They require little or no infrastructure for pre-installation in buildings, but without an external reference, errors quickly accrue [5].

In a standard PDR algorithm, the step number and length are detected using the threshold for vertical acceleration, and the accuracy of the results is not as good as that of the estimates. In practice, while walking, the velocity of one foot should be zero when it is settled down. Therefore, considering the different walking patterns of different users, the Zero-velocity information, which can improve the position result, can be used to detect the step and calculate the step length.

Meanwhile, orientation determination is also an essential issue for the PDR algorithm that requires attention because the sensor is unlikely to be axis-aligned. It continuously rotates with respect to the Navigation frame (N-frame) during the walking cycle. We must, therefore, track the rotation of the sensor by using the angular velocities provided by the gyroscopes and the magnetic field strength (MFS) from the magnetometer. However, both have their associated limitations.

Moreover, as we know, measurement errors are inevitably present within the sensor data, and a triple integration results in a potentially cubic growth in time (drift). Taking the drift into account has implications on regularly closing the integration loop by applying external constraints to the system. The most widespread PDR constraint is provided by Zero Velocity Updates (ZUPTs) [6,7,8], which is also used in this paper [9].

Hence, floor map information is important for this kind of research. To overcome the constraints we elaborated on before, a floor map can be used to further calibrate the bias and correct for unreasonable positioning results, keeping the walking trace under control. It works by rectifying the weight of the position information or correcting the heading angle, among others. For example, combining gyroscope measurements with the use of a floor map allows the orientation to be corrected using only map aids [10,11], and large heading errors are eliminated via long-range geometrical constraints exploited by Particle Filters (PFs) [2,11,12]. Unfortunately, the large number of particles makes such algorithms unrealistic in terms of real-time operations. Hence, the unscented Kalman filter, which requires a limited number of particles ($2n+1$, where n denotes the dimension of the state vector) and still confers the advantage of a Kalman filter, has been chosen for this study.

In addition, it is also well recognized that there exist many surveys on various positioning systems in the literature such as WiFi [13,14], UWB [15,16], RFID [17,18], Zigbee [19,20], and Image-Based Location (IBL) [21,22], among others. However, most of them require a fundamental facility to support their detection. In this paper, zero-velocity is detected and used to develop an improved PDR model, and thereafter, it is used to fuse floor map topology to provide a more accurate position without any other hardware preparation.

In this paper, a scheme for indoor positioning by fusing the floor map and sensor data to obtain a real-time hybrid indoor navigation result is presented. Compared with the existing technology, integrating the zero-velocity detection method and high sampling frequency could improve the accuracy and sensitivity of step detection and step length estimation to some extent. Moreover, an adaptive unscented Kalman filter (A-UKF) is presented to fuse the map information with the improved PDR result, as a compromise for considering both accuracy and Computational Load (CL). In Section 2, a type of zero-velocity detection method is proposed, and Section 3 proposes the basic theory for map matching used in this method. Subsequently, a fusion algorithm based on an A-UKF is demonstrated in Section 4. Then, Section 5 details the methodologies used in this study.Section 6 lists the equipment and environment for the experiments and the test designs to verify the accuracy and robustness of the idea, which is also analyzed. Finally, Section 7 concludes the paper.

2. AN IMPROVED PDR MODEL

2.1. Zero-Velocity Detection

The ZDET algorithm is considered the most reliable and versatile method regardless of the user and displacement patterns (walk, run, side walk, crisscross) [23]. Just as its name implies, at foot stance, the velocity is zero, while *v*, which is obtained by integrating the free acceleration, should be below some certain threshold δ. However, due to the measurement noise and bias of the accelerometers, de-noising has to be applied. The algorithm is elaborated below.

2.1.1. Acceleration Processing

As the acceleration measurements have noise and gross errors, a processing model is required before PDR model development.

1. Transform the captured accelerations. acc_i from the B-frame (B) is transformed into the Navigation frame (N) as the Figure 1 shows below using the rotation matrix:

   accNi=RB→N · accbi; then, the influence of gravity is canceled out.

2. Filter the acceleration at every stance event:

   a. Compute the accelerometer magnitude $acc\_mag = \sqrt{accX^2 + accY^2 + accZ^2}$, where accX, accY and accZ denote the tri-axis acceleration in the navigation frame.

   b. High-Pass (HP) filter accelerometer data. The Butterworth digital and analog filter, which is a kind of ordinary HP/LP filter, is used in this algorithm: *[b,a]=BUTTER(N,Wn,'s'),* where *N* is equal to 1 and "*s*" equals "high". Wnmust satisfy 0.0 < *Wn* < 1.0, with 1.0 corresponding to half the sample rate.

Then, acc_magFilt=filtfilt(b,a,acc_mag) the transfer function is [24]:

$$H(s) = \frac{b(1)s^n + b(2)s^{n-1} + \Lambda + b(n+1)}{s^n + a(2)s^n + \Lambda + a(n+1)} \tag{1}$$

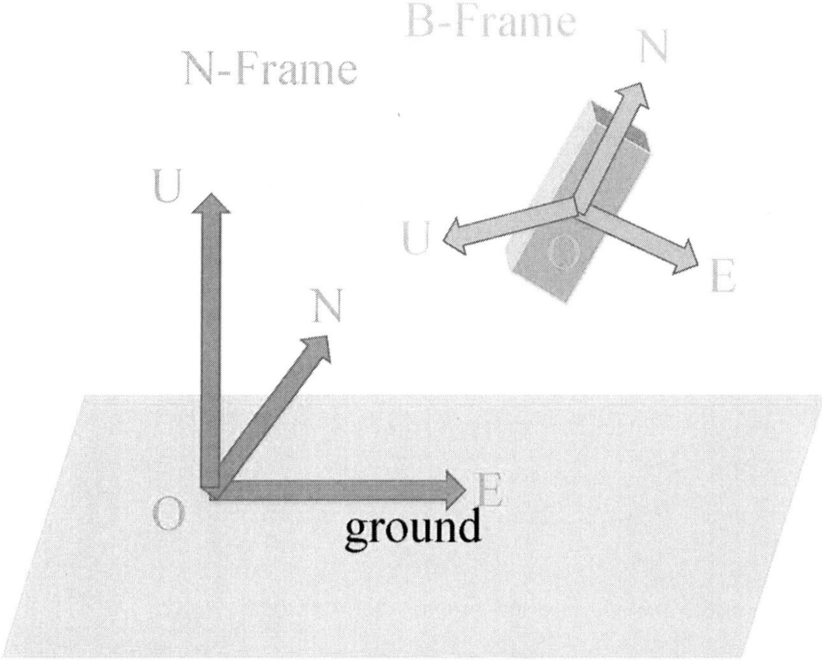

Figure 1. Illustration of the two coordinate systems.

2.1.2. Zero-Velocity Detection (ZDET) Model

We do not have a consensus on the best model for Zero-velocity detection (ZDET) for all situations. Hence, in this experiment, four kinds of detector models, provided by Nilsson [9], have been chosen as alternative schemes. They are presented below:

(1) Generalized likelihood ratio test (GLRT)

$$T = \frac{\sum_{i=1}^{n}\left(\frac{g(i)^T * g(i)}{\varepsilon_g^2} + \frac{f(i)^T * f(i)}{\varepsilon_f^2}\right)}{n} \quad (2)$$

where g(i) and f(i) represent the data on the tri-axis gyros and free acceleration, respectively, and ϵg and ϵf denote the standard deviation of the respective associated noise. Meanwhile, *n*denotes the window size of the zero-velocity detector, and *T* is the test statistic for the detector.

(2) Accelerometer measurement variance test (MV)

$$T = \frac{\sum_{i=1}^{n} \frac{(a(i)-avg)^T * (a(i)-avg)}{\varepsilon_f^2}}{n} \tag{3}$$

where *a*(*i*) denotes the raw data on the tri-axis acceleration and avg denotes the average value of the acceleration from 1 to *n*.

(3) Accelerometer measurement magnitude test (MAG)

$$T = \frac{\sum_{i=1}^{n} \frac{(norm(a(i))-grav)^2}{\varepsilon_f^2}}{n} \tag{4}$$

where *norm* is the linearization and grav denotes local gravity.

(4) Angular rate measurement energy test (ARE)

$$T = \frac{\sum_{i=1}^{n} \frac{(norm(g(i)))^2}{\varepsilon_f^2}}{n} \tag{5}$$

To verify the accuracy of the four schemes, we tested them under conditions similar to our test site (with corners and lines), and the results are shown below:

As we see in Table 1, the covariance denotes the gap between the measured result and real trajectory for each step. GLRT and ARE perform much better than the other two models. We therefore choose GLRT as our ZDET model in the experiments.

Table 1. Covariance of the four Zero-Velocity Detection (ZDET) models.

Type	Covariance (m^2)
GLRT	0.43197
MV	1.04398
MAG	2.68470
ARE	0.43931

2.2. Step Length Estimation

(1) The acceleration after step detection during each step, acc_i^N, is integrated to obtain the linear velocity velNi: velNi= velNi+accNi/SampleRate. We set the velocity to be zero when the foot is stationary [25].

(2) We obtain the position increment at step k, $\Delta P_k = (\Delta P_k(north), \Delta P_k(east), \Delta P_k(up))$, by integrating the corrected velocity vel_i^N, samples, velNi, at every single step:

$$\Delta P_k = \sum_{j=i(k-1)}^{i(k)} vel_i^N \,/\, SampleRate \tag{6}$$

(3) The 2D SL is computed by taking the horizontal Cartesian distance of the position increment:

$$SL_{ZUPT_k} = \sqrt{\Delta P_k (north)^2 + \Delta P_k (east)^2} \tag{7}$$

The step length estimation results for the respective Walking and Running models are shown in Table 2:

Table 2. Gait recognition based on the ZDET algorithm.

Walking	Step Number	Accuracy Rate	Running	Step Number	Accuracy Rate
1	50	100.0%	1	10	100.0%
2	60	100.0%	2	20	100.0%
3	70	97.1%	3	30	96.7%
4	80	92.5%	4	40	92.5%
5	90	96.7%	5	50	98.0%
6	100	96.0%	6	60	86.7%

The results demonstrate that the accuracy of Walking and Running for several tests with the same experimenter, but different steps, are above 90% and 85%, respectively (compared with ground truth), indicating that the method developed is a reliable method for indoor experiments.

2.3. Heading Determination by Fusing Gyroscope and Magnetometer Data

Heading determination is a significant component of PDR-based positioning. The heading angle ψ is defined as the angle of rotation about the Z-axis with respect to the horizon/ground, which can be estimated using a gyroscope integrated with a magnetometer as well as by a magnetic field.

First, we fuse the data from the magnetometer and gyroscope using the heading estimation algorithm presented by Wonho Kang in [21]. The fused heading angle is calculated as follows:

$$\begin{cases} \theta_k = \alpha\theta_{k-1} + \beta\theta_{m,k} + \gamma\theta_{g,k}, & \theta_{\Delta,c} \leq \theta_c, \theta_{\Delta,m} \leq \theta_m \\ \theta_k = \beta\theta_{m,k} + \gamma\theta_{g,k}, & \theta_{\Delta,c} \leq \theta_c, \theta_{\Delta,m} > \theta_m \\ \theta_k = \alpha\theta_{k-1}, & \theta_{\Delta,c} > \theta_c, \theta_{\Delta,m} \leq \theta_m \\ \theta_k = \alpha\theta_{k-1} + \gamma\theta_{g,k}, & \theta_{\Delta,c} > \theta_c, \theta_{\Delta,m} > \theta_m \end{cases} \tag{8}$$

$$\theta_{\Delta,c} = \left|\theta_{m,k} - \theta_{g,k}\right|, \quad \theta_{\Delta,m} = \left|\theta_{m,k} - \theta_{m,k-1}\right|.$$

where *m* and *n* denote the magnetometer and the gyroscope, respectively, and *α*, *β* and *γ* are the weights of the current measurements from the gyroscope and the magnetometer. θm,k and θg,kdenote the measurements acquired by the gyroscope and the magnetometer, respectively, for the kth step. θm is the standard deviation of the magnetometer, and θc is the correlation between the magnetometer and the gyroscope readings. θΔ,c is the difference between θm,k and θg,k. θΔ,m is the difference in the magnetometer readings between two consecutive steps *k* and *k* – *1*.

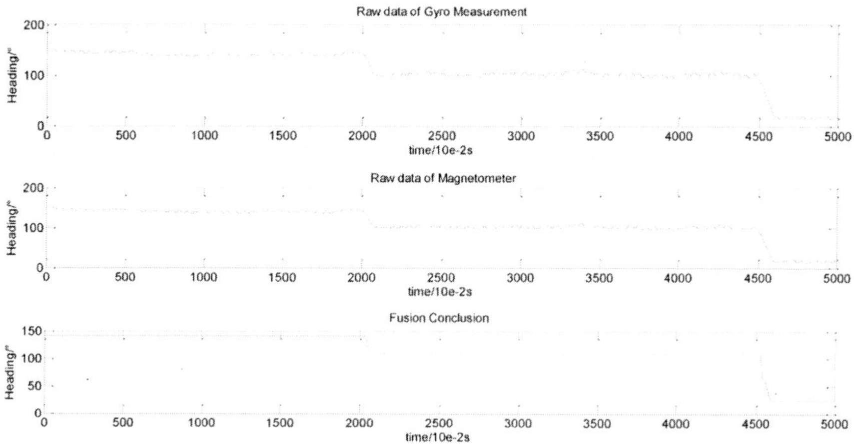

Figure 2. A test of heading fusion.

Figure 2 presents the fusion results of a test for which a person walked forward dozens of steps with a smartphone held firmly in his hand and then crossed the first corner (turn of almost 38.5 degrees). After a long corridor, the experimenter eventually turned left (turn of approximately 90 degrees) and reach the destination. The results reveal that the pseudo-heading measurements recorded by the magnetometer before and after the turn are considerably smoothed by the proposed algorithm, reducing fluctuation to a certain extent [12].

2.4. The Improved PDR

Presented below is the flow chart (Figure 3) for the improved PDR method:

Compared with the traditional PDR method, ZDET could render the velocity to be zero, which eliminates the residual error in the velocity.

3. FLOOR MAP MATCHING ALGORITHM

Map matching is a kind of position-calibrated method based on software technology, which is used as a sort of pseudo-measured technology. The main idea involves comparing localization results and the passable zone in the floor map, from which the most suitable trajectory should be found. We can use the Point of Interest (POI), such as a corner, to calibrate the accumulated error. In a broad sense, map matching in Indoor Positioning can be classified into a combined application category rather than a framework model category [26].

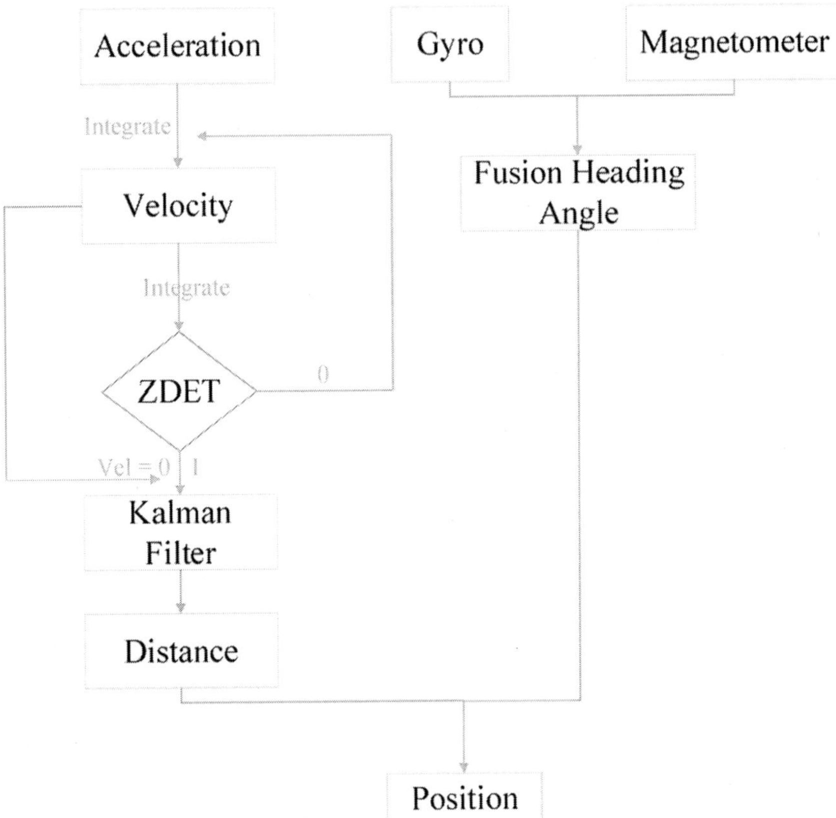

Figure 3. Flow chart of the improved PDR.

3.1. Correlation Matching Algorithm Based on Projection (CMAP)

The CMAP is a kind of matching algorithm that involves mapping points to arcs. As shown inFigure 4 when the user's position (Point *P*) exceeds the set buffer zone, by comparing the respective distances (d_1, d_2 and d_3) between Point *P* and each Line (L_1, L_2 and L_3), an optimal solution that matches the user's position to the nearest lines instead of knots and sharp points ("nearest" here represents the shortest distance between the terminal and road) can be determined. In most situations, this strategy performs better than the point to point (P2P) method [10].

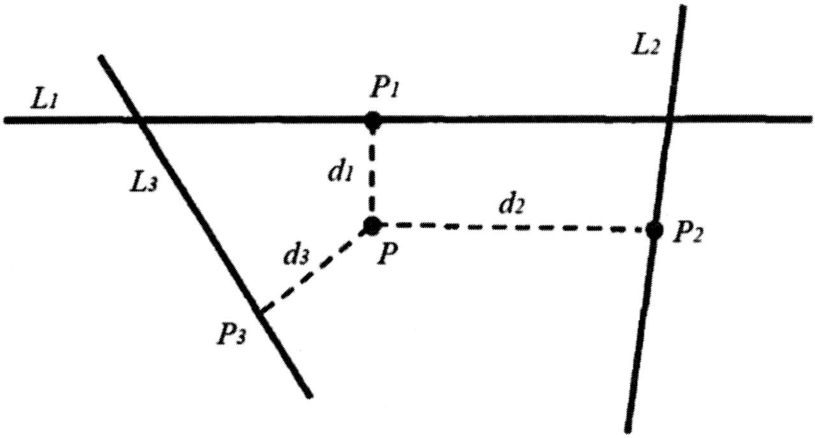

Figure 4. Demonstration of Correlation Matching Algorithm based on Projection (CMAP).

3.2. Zone Division

First, the base map is to be divided into several components, such as passable zones and impassable zones. The passable zone can also be divided into linear vector zones and non-linear vector zones [12].

As the width of the corridor (indoor) has a limit, the walking trace can be seen as an approximate straight line, denoted by the grey components in Figure 5, and the corners can be thought of as non-linear zones.

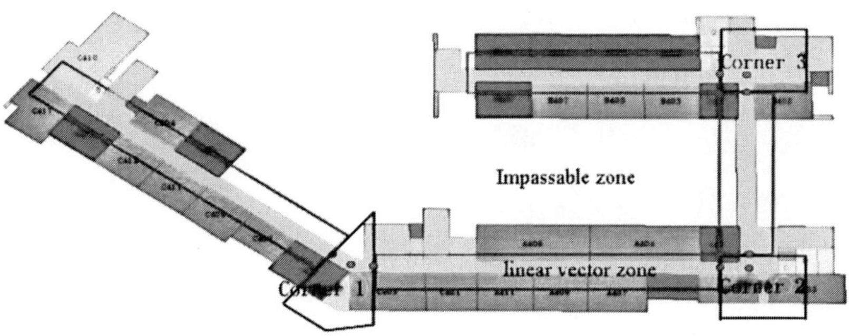

Figure 5. Zone division.

3.3. Vector Matching

When the walking trace is out of range, such as observed when running into an impassable zone, it can be rectified by CMAP to a certain extent.

Moreover, the POI can be evaluated using the accumulated gyroscope data, as shown in Figure 6. When we turn a corner, the valley of accumulated data may be under a certain limit (10 rad in this instance).

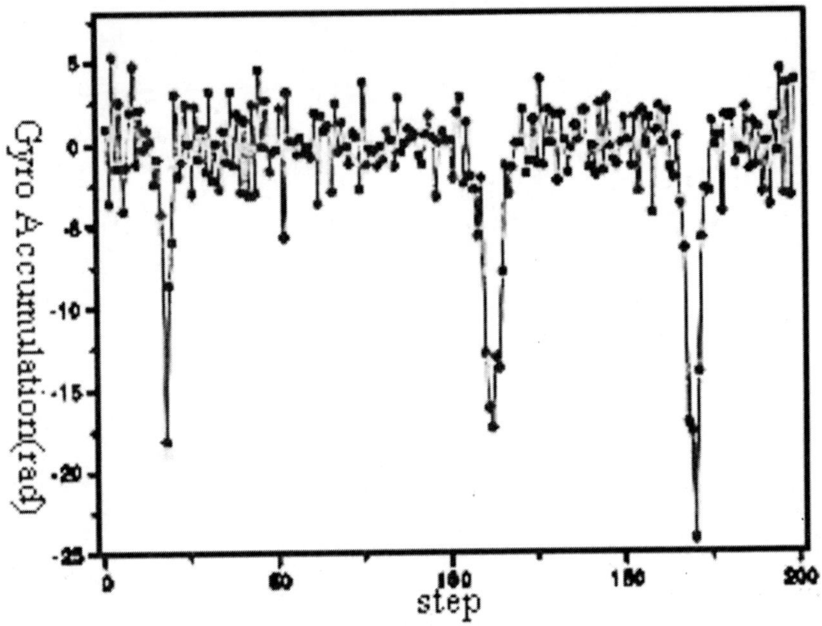

Figure 6. Accumulated gyroscope data.

The method for floor map matching (Figure 7) is shown below:

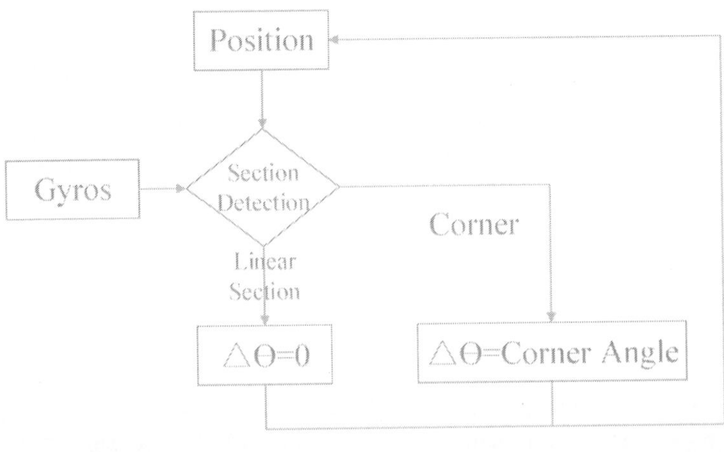

Figure 7. The flow of map matching.

4. ADAPTIVE UNSCENTED KALMAN FILTER (A-UKF)

Despite the better accuracy of the approaches described above, the absolute heading angle obtained using pre-process formula (5) may exhibit unpredictable errors, which could be eliminated by map matching (CMAP) to a certain extent. Considering the combination of the above methods and the non-linearized features of the observation equation, the Unscented Kalman Filter may be a suitable method to use [27,28,29].

4.1. Dynamic Equation

The system state is estimated using recursive UKF equations, and the state equation is

$$X_k = \begin{pmatrix} N_k \\ E_k \\ \theta_k \end{pmatrix} = \begin{pmatrix} N_{k-1} + \tilde{s} \cdot \cos\theta_{k-1} \\ E_{k-1} + \tilde{s} \cdot \sin\theta_{k-1} \\ \theta_{k-1} + \tilde{\theta} \end{pmatrix} + W_{k-1} \tag{9}$$

where Nk and Ek are the position states in the northern direction and the eastern direction, respectively, at time *k*; θk denotes the heading angle of the pedestrian; s˜ and θ˜ denote the step length and angle velocity, respectively; and W_{k-1} is the system noise at time *k* – *1*.

4.2. Observation Equation

The observation equation for the integrated step length based on AHRS, the heading angle and orientation measurements can be written as follows:

$$Z_k = \begin{bmatrix} s_k \\ \Delta\theta_k \\ \theta_k \end{bmatrix} = \begin{bmatrix} \sqrt{(x_k - x_{k-1})^2 + (y_k - y_{k-1})^2} \\ \theta_k - \theta_{k-1} \\ \theta_k \end{bmatrix} + V \tag{10}$$

where sk is the step length calculated using velocity increments of two consecutive steps and Δθkis the heading angle variable determined by CMAP, including corners and galleries. θk is the k^{th}heading angle measured by formula (5), and *V* is the measurement noise vector.

4.3. A-UKF Algorithm

If the dynamic model of the system is abnormal, the variance of the dynamic model has to be adjusted to control the parameter estimation. Based on an Unscented Kalman filter (UKF) and the adaptive filtering algorithm, using the UKF algorithm for importance sampling, the sampling variance and mean value can be calculated by establishing the adaptive factor and real-time adjustment of the sampling variance, thereby weakening the error state model [30].

The steps for calculating the adaptive state parameter variance, which is different from ordinary UKF, is as follows:

(1) Adaptive parameter

The distinguishing statistic of the adaptive factor is the discrepancy of the state. $\bar{X}_{k|k-1}$ denotes the state forecasting value, $\tilde{x}_k$ denotes the robust state solution, and $\bar{V}_k$ denotes the difference between the state forecasting value and robust solution. When $\bar{V}_k$ reaches a certain limit, the system is considered to be in an abnormal state [31].

$$\bar{V}_k = \bar{x}_{k|k-1} - x_k \tag{11}$$

$$\Delta\bar{V}_k = \frac{\bar{V}_k}{\sqrt{tr(P_{\bar{V}_k})}} \tag{12}$$

where tr(·) denotes the trace formula. The form of the adaptive factor function is:

$$\gamma_k = \begin{cases} 1 & |\Delta\bar{V}_k| \le c \\ \dfrac{c}{|\Delta\bar{V}_k|} & |\Delta\bar{V}_k| > c \end{cases} \tag{13}$$

where C is constant and ranges in value between 0.85 and 1.

(2) Calculating the adaptive filter equivalent variance

$$\overline{P}_{vv} = (P_{vv} - P_{v_k}) / \gamma_k + P_{v_k} \tag{14}$$

$$\overline{K}_k = P_{xv} \overline{P}_{vv} / \gamma_k \tag{15}$$

$$P_k = P_{k|k-1} / \gamma_k - \overline{K}_k P_{vv} \overline{K}_k^T \tag{16}$$

where $P_{k|k-1}$ denotes the variance of the Predictive Sample Point (PSP) and Pvv and Pxvrepresent the variance and covariance of the Predictive Measurement Point (PMP), respectively.

The flowchart is shown below in Figure 8:

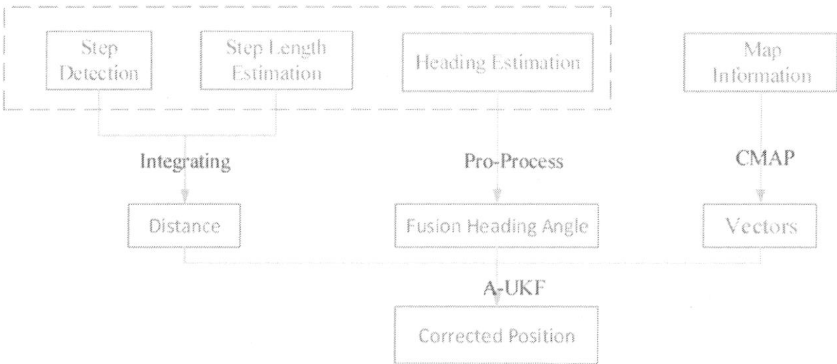

Figure 8. Flowchart of the filter process.

4.4. Adaptive Factor Test

To verify the effectiveness of the adaptive factor, an experiment was designed for proving the effect of A-UKF. The route in the SESSI building is shown below in Figure 9:

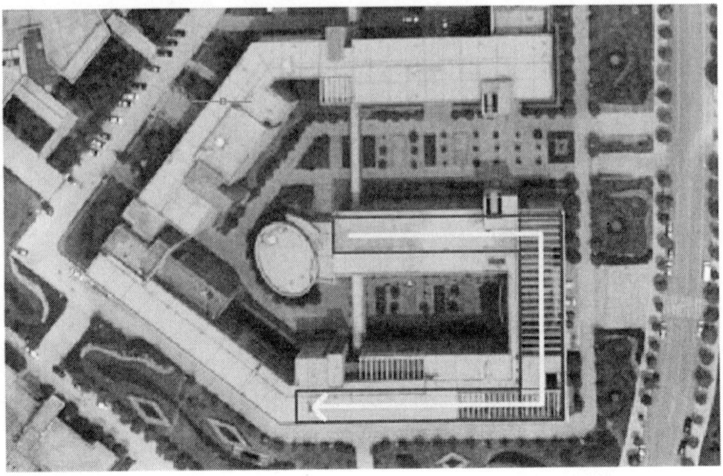

Figure 9. Test Route in the School of Environmental Science and Spatial Informatics (SESSI) Building.

Moreover, three artificial error points are added into the data for each linear corridor in the route (approximately 60 steps, 126 steps and 219 steps), and the results are shown below:

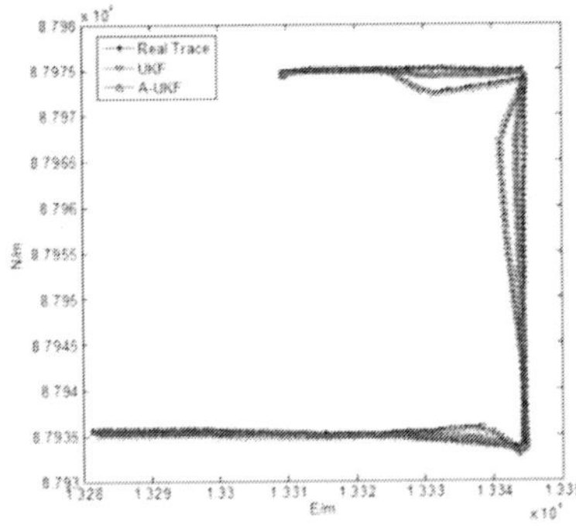

(a)

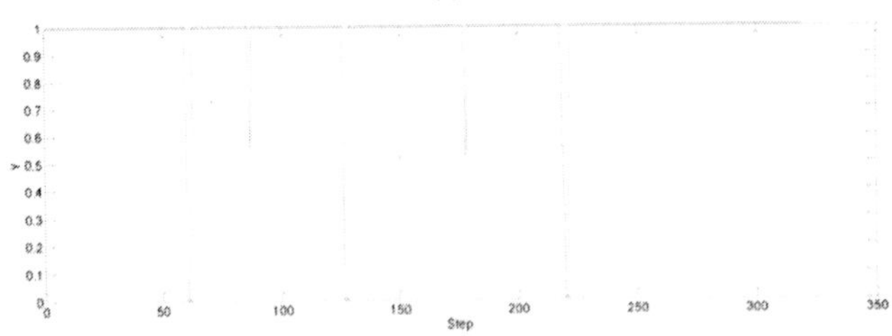

(b)

Figure 10. The results of the test: (**a**) estimated position based on Unscented Kalman Filter (UKF) and A-UKF and (**b**) variance of the adaptive parameter.

As we can see in Figure 10a, the A-UKF trace is much closer to the real trace than the classic UKF trace, and it eliminates the influence of artificial error more rapidly. In addition, from Figure 10b, we see that when a gross error appears, the adaptive factor will rectify it, as does the two corners method, but with much less fluctuation, which means that the corners may cause some error that is still smaller than the gross error, which is also corrected.

5. AN IMPROVED PDR/MAGNETOMETER/FLOOR MAP INTEGRATION SCHEME

In summary, the methodologies used are described below:

1. Theoretical analysis. A series of basic studies have been analyzed. As mentioned in the introduction, recent PDR systems are becoming a feasible option for indoor localization due to the continual miniaturization of inertial sensors (Inertial Management Unit and Attitude Heading Reference System). Moreover, based on the three main components of PDR (step detection, step length estimation and orientation determination), we select the ZDET algorithm, an unusual but accurate algorithm, to detect the step. Then, during every step, a continuous integration algorithm is used to determine the distance with the acceleration, which is recorded from the accelerometer and transformed to N-frame coordinates. Moreover, the determination of orientation should be the most essential step. We therefore use a hybrid filter to fuse the data, provided by the AHRS system, to guarantee the stability and precision of the orientation.
2. Integration Methodology Development. In the experiment, several filters are tested, such as the Extended Kalman Filter, Unscented Kalman Filter, Particle Filter, and others. Considering various factors, including Measuring Accuracy (MA) and Computational Load (CL), we eventually choose one kind of adaptive unscented Kalman filter (A-UKF) as the fusion algorithm.
3. Physical System Implementation and Tests. A specific flowchart is shown in Figure 11:

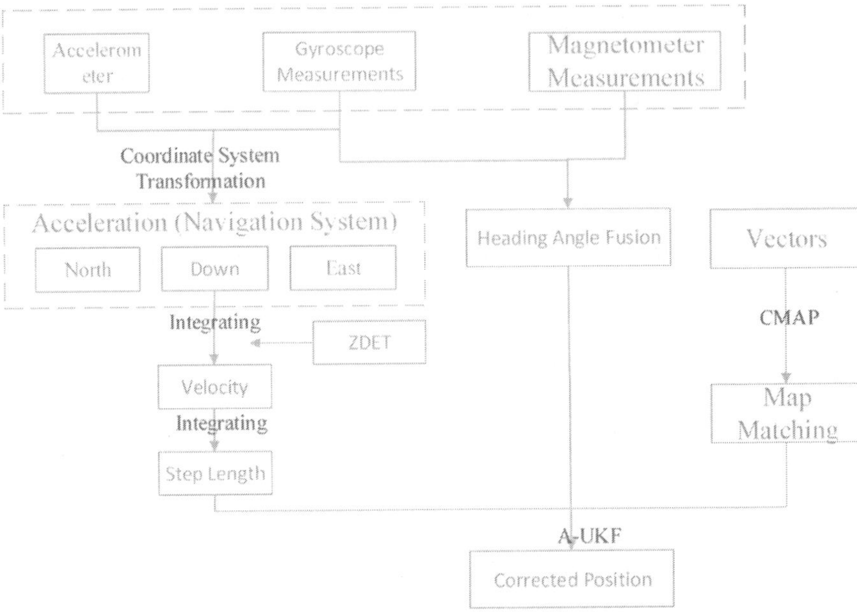

Figure 11. The general flowchart.

6. EXPERIMENT AND ANALYSIS

6.1. Equipment and Situation

We use a commercially available Inertial Measurement Unit, x-IMU, from x-io Technologies in UK. Figure 12 shows this sensor. It is 55 × 35 × 18 mm (L × W × H) in size and almost 50 g in weight.

Figure 12. Photos of the equipment and their usage: (**a**) the x-IMU and (**b**) foot-mount installation.

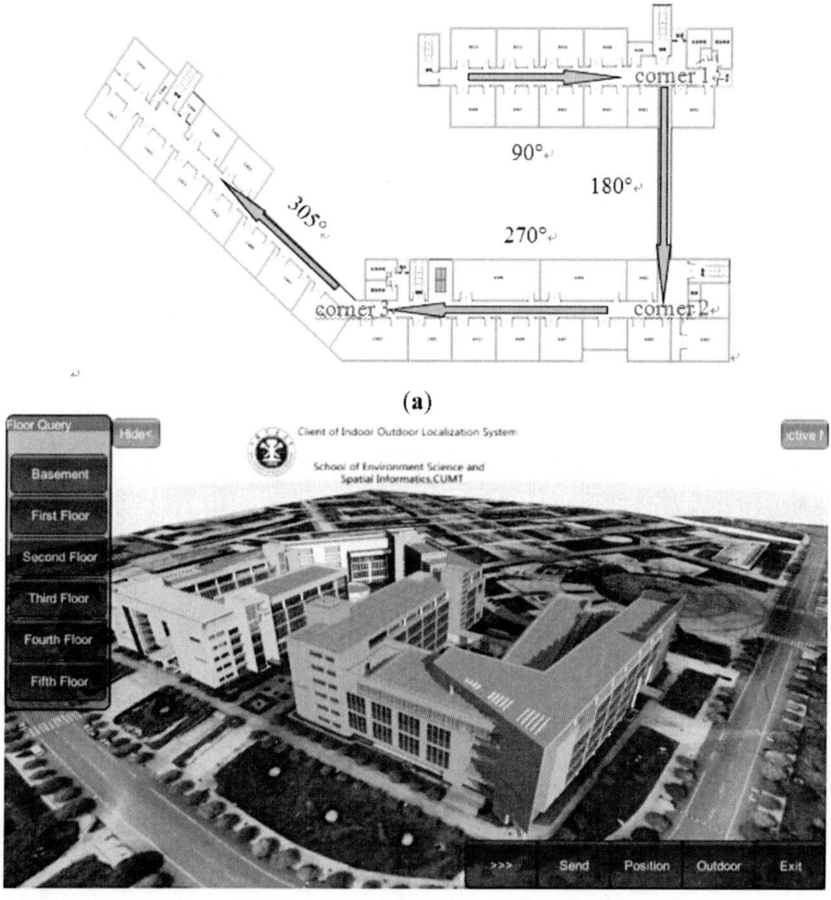

Figure 13. Experimental site: (**a**) floor map of the fourth floor and (**b**) 3D model of the experimental site.

The x-IMU was designed to be the most versatile Inertial Measurement Unit (IMU) and Attitude Heading Reference System (AHRS) platform available. It hosts on-board sensors, a configurable auxiliary port and real-time communication via USB, Bluetooth or UART, which makes it both a powerful sensor and controller. The on-board SD card, battery charger (*via* USB), real-time clock/calendar and motion trigger wake up also make the x-IMU an ideal standalone data logger.

The x-IMU has three orthogonally oriented accelerometers, three gyroscopes and three magnetometers. The accelerometers and gyroscopes are MEMS solid state with capacitive readouts, providing linear acceleration and rate of turn, respectively. Magnetometers use a thin-film magnetoresistive principle to measure the Earth's magnetic field.

The performance of each individual MEMS sensor within this facility is summarized in Table 1. They suffer from a significant bias, and this bias also varies over time; therefore, PDR algorithms have the challenge of avoiding excessive error accumulation (drift) during integration.

To verify the effectiveness of the proposed algorithm, a field experiment was performed on the fourth floor of the School of Environmental Science and Spatial Informatics (SESSI) building on the campus of China University of Mining and Technology (CUMT) in Xuzhou, Jiangsu, China. In Figure 13a, blue arrows indicate linear corridors, and red "corners" indicate non-linear corridors and other areas that are unavailable for a person to pass through. A type of AHRS named x-imu produced by x-io Technologies Limited was used as the user terminal in the experiment; its technical specifications are shown in Table 3.

Table 3. Technical specifications of the IMU.

Instrument	Scale Factor	Random Walk
Accelerometer (m/s^2)	10.3333	±0.0005
Gyroscope (m/s^2)	5.2457	±0.0003
Magnetometer (μT)	800.0000	±0.0400

6.2. Experiment

To compare the tracking results, four schemes were designed:

Scheme 1: Floor-map-aided PDR-based tracking result by hand-held smartphone (MPDR).

Scheme 2: ZDET-based indoor positioning test with ordinary data (ZDET).

Scheme 3: ZDET-based indoor positioning test with map-matching (MZDET).

Map matching (achieved by A-UKF) was used to correct the information from different sensors for Schemes 1 and 3. In Schemes 2 and 3, the ZDET was used for step detection and step length estimation, and the ordinary PDR was accepted by Scheme 1. The floor level should, in general, be determined in advance from Z-axis accelerometer measurements, although only 2D positioning was considered here.

Figure 14 shows the position trajectories for the three schemes. The colorful lines represent the travel paths calculated using the different schemes. In the test, the test participant walked with a uniform and stable gait; therefore, the true trajectory should be stable.

As seen from Figure 15, the map-matching algorithm could help PDR eliminate the accumulated error, which should be an inevitable problem for classic PDR. However, some issues are still revealed to some degree; the step length estimation and step detection may be imprecise, and the error will be accumulated just as a black line in a block. Moreover, compared with ZDET, the positioning results of the MZDET are more accurate and rarely suffer from the influence of a variable heading angle.

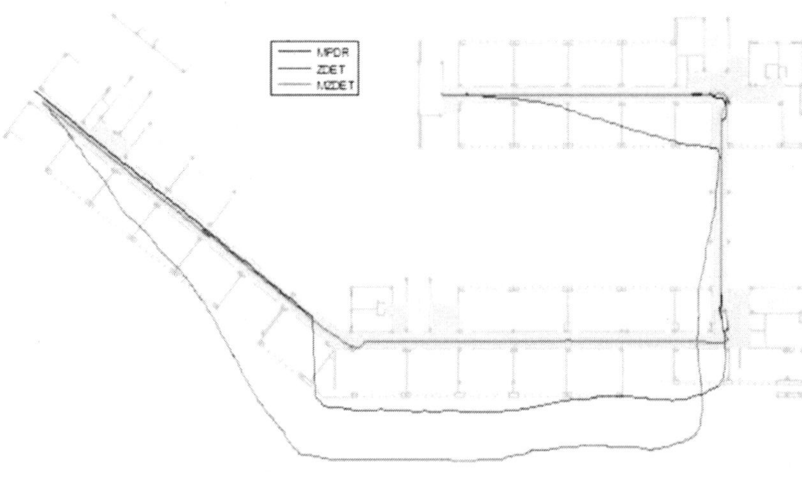

Figure 14. Estimated position of the three schemes.

Moreover, compared with an ordinary PDR algorithm, the figure below illustrates that the UKF algorithm preserves the continuity and stability of the PDR algorithm while simultaneously restricting the accumulation of errors, thereby improving the positioning accuracy.

Figure 16 shows the step series of the position errors of the three schemes with respect to the reference positions provided by a master station. The largest errors of the MZDET integration algorithm were observed during hundreds of steps, which it shows much better than the other two methods.

Moreover, we can see in Figure 17 that when a gross error (huge error) is observed, the adaptive factor is able to react and rectifies it rapidly.

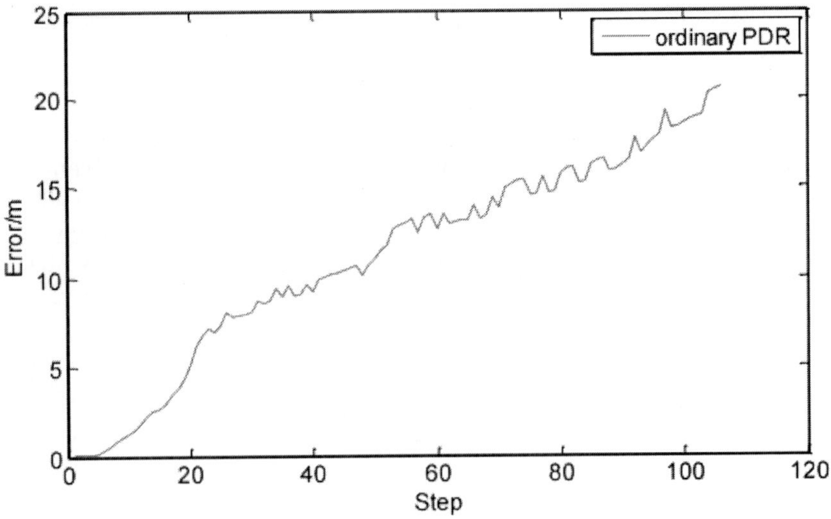

Figure 15. The accumulated error of the PDR algorithm.

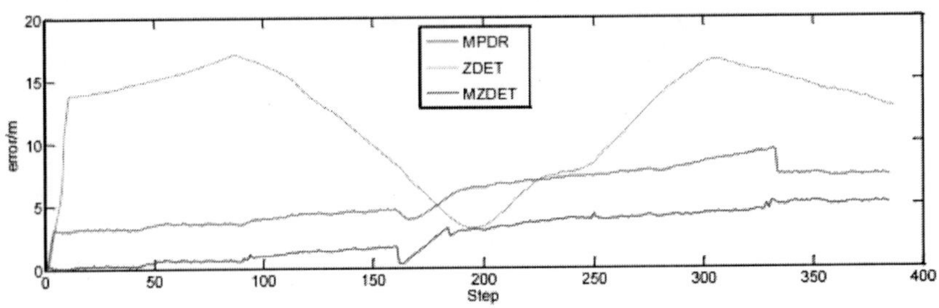

Figure 16. The error step series of the three schemes.

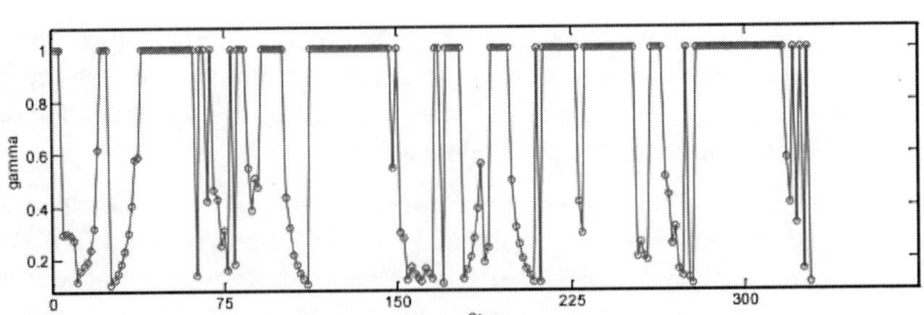

Figure 17. The variance of the adaptive parameter.

The MSE, the average error (AE) and the maximum error (ME) reveal that the map-aided ZDET algorithm achieved the most reliable and accurate positioning results. In this test, the accuracy indicated by the MSE of the integrated MZDET indoor positioning algorithm was improved by 78.3% compared with the pure ZDET algorithm, and the AE and ME were reduced by 73.8% and 69.2%, respectively. Compared with the MPDR algorithm, the MSE of the integrated WPO algorithm was improved by 32.2%, and the AE and ME were reduced by 39.5% and 53.7%, respectively (Table 4).

Table 4. Error analysis of the three schemes.

ERROR	MPDR	ZDET	MZDET
MSE/m	4.443	13.903	3.011
AE/m	4.360	12.190	2.639
ME/m	11.427	17.142	5.285

As seen from the comparison, the integrated MZDET positioning algorithm results in an improvement in the accuracy, reliability, and calculation rate and a decrease in the accumulated error; therefore, this method is superior to the others to a certain extent.

6.3. Robustness Test

To demonstrate the robustness of the A-UKF integration algorithm for a foot-mounted indoor positioning system, some gross errors (three times the threshold of the step length covariance) were added to the step length observations as listed in Table 5:

Table 5. Gross errors added to the step length measurements.

Time (s)	30	60	85	126	178	219	269
s (m)	3	3	3	3	3	3	3

Figure 18 illustrates the positioning trajectories determined using the UKF algorithm and the A-UKF algorithm. As shown in Table 6, seven gross-error-contaminated points, which cause the trajectory to be deteriorated, are observed around each corner and linear corridor in Figure 13a.Table 6 shows the residuals of these two algorithms for each affected time point (as shown in the table before), and the results demonstrate that the A-UKF algorithm achieves a much higher reliability and accuracy.

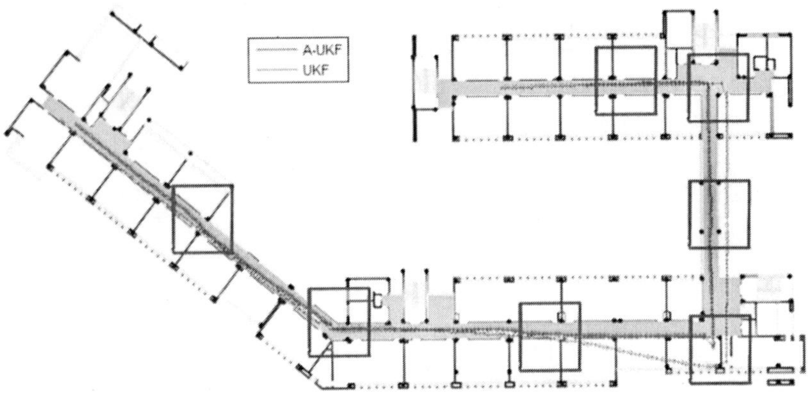

Figure 18. Estimated position based on A-UKF and UKF.

Table 6. Residuals of the gross-error epochs.

Time (s)	UKF(m)	A-UKF (m)
30	1.50	1.39
60	1.99	0.43
93	3.00	0.32
126	2.57	0.58
172	1.84	0.76
219	1.48	1.08
269	1.50	1.05

Gross errors were also added to the θ_k measurements (three times the threshold of the heading angle covariance), as listed in Table 7, to verify the robustness of both algorithms. The resulting positioning trajectories obtained by using the UKF algorithm and the A-UKF algorithm are shown in Figure 19. As shown in Figure 19, the seven gross-error-contaminated points cause the trajectory to be deteriorated around the affected epochs as a result of the algorithms' recursions. Table 8 lists the residuals of the gross-error-contaminated epochs.

Table 7. Gross errors added to the orientation measurements.

Time (s)	30	60	93	126	172	219	269
θ_k (°)	+20	+20	+20	+20	+20	+20	+20

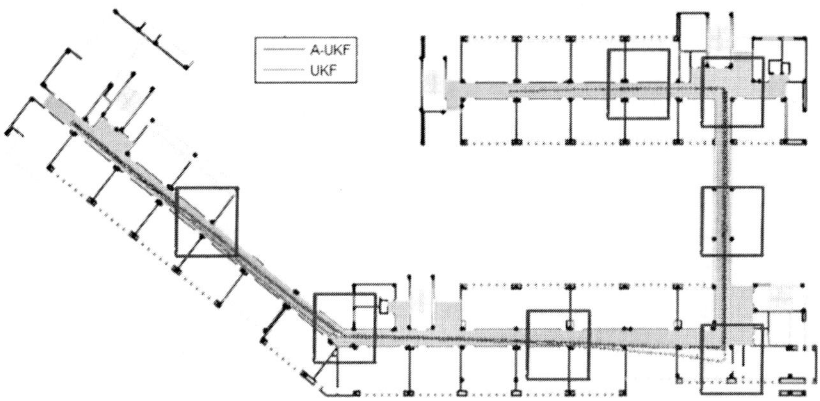

Figure 19. Estimated position based on UKF and A-UKF.

Table 8. Residuals of the gross-error epochs.

Time (s)	UKF (m)	A-UKF (m)
30	0.23	2.13
60	1.08	1.10
93	0.92	0.14
126	2.15	0.27
172	1.78	0.50
219	0.56	0.46
269	1.08	0.75

These results demonstrate that the A-UKF algorithm is more robust than the ordinary UKF algorithm, regardless of where the gross error originated.

These observations yield the following conclusions regarding the investigated positioning algorithms for indoor navigation systems.

1. Considering the issue of step detection, a ZDET method was proposed to improve the accuracy of detection, thereby decreasing the fluctuation to an acceptable level and improving the accuracy of the entire algorithm.
2. Despite the high accuracy provided by zero-velocity detection, PDR always requires a considerable precision of orientation, and therefore, most experiments using PDR require constraints to cope with the information obtained from the gyroscope or magnetometer. However, incorporating an electronic map of the structure by UKF into the analysis eliminates redundant operations, thereby helping to improve the computational speed and quality.
3. Considering the stability of UKF, the adaptive parameter is integrated into the algorithm. The experiments show that compared with the classic UKF, A-UKF offers a much more powerful robustness, improving the positioning reliability and producing a dramatic upward shift in operating rate and quality.
4. In view of the considerable CL of A-UKF, which could load upon read-time calculation on an ordinary smartphone, there is the possibility of quasi-real-time measurements using mobile phones instead of a central computer, as has generally been used in other recent experiments.

7. CONCLUSIONS

This paper investigated several positioning algorithms for indoor navigation systems. A Zero-Velocity Update (ZDET) algorithm step detection method was presented. The proposed floor-map-aided ZDET algorithm was combined with an adaptive unscented Kalman filter (A-UKF). The experimental test results indicate that the ZDET methods eliminated the error of step length estimation and step detection to a certain extent. It was also demonstrated that the A-UKF could be very useful for correcting the heading angle and improving the quality and accuracy of the positioning results. Further development of the algorithm will focus on fusing other data sources, such as WiFi and ibeacon based on Bluetooth Low Energy (BLE), which could improve the accuracy and stability of the algorithm and make it more efficient.

ACKNOWLEDGMENTS

This work was partially supported by Special funds for basic scientific research business of Central University (No. 2015XKMS051).

AUTHOR CONTRIBUTIONS

The corresponding author Jian Wang proposed the research, organized the equipment and guide writing the paper. Andong Hu purposed the experimental program, collected most of the field data, did the analysis and drafted the manuscript. Xin Li and Yan Wang performed part of the field data and involved in drafting the manuscript.

REFERENCES

1. Cho, S.Y.; Park, C.G. Mems based pedestrian navigation system. *J. Navig.* **2006**, *59*, 135–153.
2. Klepal, M.; Beauregard, S. A backtracking particle filter for fusing building plans with PDR displacement estimates. In Proceedings of 5th Workshop on Positioning, Navigation and Communication, Hannover, Germany, 27 March 2008; pp. 207–212.
3. Li, F.; Zhao, C.; Ding, G.; Gong, J.; Liu, C.; Zhao, F. A reliable and accurate indoor localization method using phone inertial sensors. In Proceedings of the 14th International Conference on Ubiquitous Computing, Pittsburgh, PA, USA, 5–8 September 2012; pp. 421–430.
4. Kourogi, M.; Kurata, T. Personal positioning based on walking locomotion analysis with self-contained sensors and a wearable camera. In Proceedings of the 2nd IEEE/ACM International Symposium on Mixed and Augmented Reality, Washington, DC, USA, 7–10 October 2003; pp. 103–112.
5. Harle, R. A survey of indoor inertial positioning systems for pedestrians. *IEEE Commun. Surv. Tutor.* **2013**, *15*, 1281–1293.
6. Lan, K.C.; Shih, W.Y. Using smart-phones and floor plans for indoor location tracking. *IEEE Trans. Hum. Mach. Syst.* **2014**, *44*, 211–221.
7. Colomar, D.S.; Nilsson, J.; Handel, P. Smoothing for ZUPT-aided INSs. In Proceedings of 2012 International Conference on Indoor Positioning and Indoor Navigation (IPIN), Sydney, Australia, 13–15 November 2012; pp. 1–5.
8. Nilsson, J.O.; Skogl, I.; Handel, P.; Hari, K.V.S. Foot-mounted INS for everybody—An open-source embedded implementation. In Proceedings of the 2012 IEEE/ION Position Location and Navigation Symposium (PLANS), Myrtle Beach, SC, USA, 23–26 April 2012; pp. 140–145.

9. Handel, P.; Rantakokko, J.; Skog, I.; Nilsson, J.O. Zero-velocity detection—An algorithm evaluation. *IEEE Trans. Biomed. Eng.* **2010**, *57*, 2657–2666.

10. Attia, M.; Moussa, A.; El-Sheimy, N. Map aided pedestrian dead reckoning using buildings information for indoor navigation applications. *Positioning* **2013**, *4*, 227–239.

11. Nurminen, H.; Ristimaki, A.; Ali-Loytty, S.; Piché, R. Particle filter and smoother for indoor localization. In Proceedings of 2013 International Conference on Indoor Positioning and Indoor Navigation (IPIN), Montleliard-Belfort, France, 28–31 October 2013; pp. 1–10.

12. Wang, J.; Hu, A.; Liu, C.; Li, X. A floor-map-aided WIFI/pseudo-odometry integration algorithm for an indoor positioning system. *Sensors* **2015**, *15*, 7096–7124.

13. Lim, C.-H.; Wan, Y.; Ng, B.-P.; See, C. A real-time indoor WIFI localization system utilizing smart antennas. *IEEE Trans. Consum. Electron.* **2007**, *53*, 618–622.

14. Lim, J.-S.; Jang, W.-H.; Yoon, G.-W.; Han, D.-S. Radio map update automation for WIFI positioning systems. *IEEE Commun. Lett.* **2013**, *17*, 693–696.

15. M.R.; Kuhn, M.J.; To, G.; Fathy, A.E. Integration of UWB and wireless pressure mapping in surgical navigation. *IEEE Trans. Microw. Theory Tech.* **2009**, *57*, 2550–2564.

16. Marano, S.; Gifford, W.M.; Wymeersch, H.; Win, M.Z. Nlos identification and mitigation for localization based on UWB experimental data. *IEEE J. Sel. Areas Commun.* **2010**, *28*, 1026–1035.

17. Sanpechuda, T.; Kovavisaruch, L. A review of RFID localization: Applications and techniques. In Proceedings of the 5th International Conference on Electrical Engineering/Electronics, Computer, Telecommunications and Information Technology, Krabi, Thailand, 14–17 May 2008; pp. 769–772.

18. Kim, T.; Shin, J.; Tak, S. Cell planning for indoor object tracking based on RFID. In Proceedings of the 10th International Conference on Mobile Data Management: Systems, Services and Middleware, Taipei, Taiwan, 18–20 May 2009; pp. 709–713.

19. Nabih, A.K.; Osman, H.; Gomaa, M.; Aly, G.M. New fuzzy-based indoor positioning scheme using ZigBee wireless protocol. In Proceedings of the 7th International Conference on Computer Engineering & Systems (ICCES), Cairo, Egypt, 27–29 November 2012; pp. 16–20.

20. Chang, A.Y.; Liu, T.-C. Performance evaluation of real-time indoor positioning with active RFID and ZigBee-based WSN systems. *J. Vibroeng.* **2013**, *15*, 736.

21. Ghinamo, G.; Corbi, C.; Lovisolo, P.; Lingua, A.; Aicardi, I.; Grasso, N. Accurate positioning and orientation estimation in urban environment based on 3D models. In *New Trends in Image Analysis and Processing—ICIAP 2015 Workshops*; Murino, V., Ed.; Springer: Cham, Switzerland, 2015; pp. 185–192.

22. Ghinamo, G.; Corbi, C.; Francini, G.; Lepsoy, S.; Lovisolo, P.; Lingua, A.; Aicardi, I. The MPEG7 visual search solution for image recognition based positioning using 3D models. In Proceedings of the 27th International Technical Meeting of The Satellite Division of the Institute of Navigation, Tampa, FL, USA, 8–12 September 2014.

23. Feliz Alonso, R.; Zalama Casanova, E.; Gómez García-Bermejo, J. Pedestrian tracking using inertial sensors. *J. Phys. Agents* **2009**, *3*, 35–43.

24. Madgwick, S.O. An Efficient Orientation Filter for Inertial and Inertial/Magnetic Sensor Arrays. Available online: https://www.samba.org/tridge/UAV/madgwick_internal_report.pdf (accessed on 30 April 2010).

25. Madgwick, S.O.H.; Harrison, A.J.L.; Vaidyanathan, A. Estimation of IMU and MARG orientation using a gradient descent algorithm. In Proceedings of the 2011 IEEE International Conference on Rehabilitation Robotics (ICORR), Zurich, Switzerland, 29 June–1 July 2011; pp. 1–7.

26. Bernstein, D.; Kornhauser, A. *An Introduction to Map Matching for Personal Navigation Assistants*; New Jersey TIDE Center: Newark, NJ, USA, 1998.

27. Jiménez, A.R.; Seco, F.; Prieto, J.C.; Guevara, J. Indoor pedestrian navigation using an INS/EKF framework for yaw drift reduction and a foot-mounted IMU. In Proceedings of the 7th Workshop on Positioning Navigation and Communication (WPNC), Dresden, Germany, 11–12 March 2010; pp. 135–143.

28. Moder, T.; Hafner, P.; Wisiol, K.; Wieser, M. 3D indoor positioning with pedestrian dead reckoning and activity recognition based on bayes filtering. In Proceedings of the International Conference on Indoor Positioning and Indoor Navigation, Busan, Korea, 27–30 October 2014.

29. Zampella, F.; Khider, M.; Robertson, P.; Jiménez, A. Unscented kalman filter and magnetic angular rate update (MARU) for an

improved pedestrian dead-reckoning. In Proceedings of the 2012 IEEE/ION Position Location and Navigation Symposium (PLANS), Myrtle Beach, SC, USA, 21–23 September 2011; pp. 129–139.

30. Yang, Y.; Gao, W. An optimal adaptive kalman filter. *J. Geod.* **2006**, *80*, 177–183.

31. Ren, X.; Yang, Y. Robust spectral analysis of time series of Geo ionospheric correction. *J. Geod. Geodyn.* **2014**, *2*, 115–119.

Chapter 1

Adaptive "Cubature and Sigma Points" Kalman Filtering Applied to MEMS IMU/GNSS Data Fusion during Measurement Outlier

Hamza Benzerrouk[1], Hassen Salhi[1], Alexander Nebylov[2]

[1]SET Laboratory (Systèmes Electriques et Télecommande) of Electronic Department of Saad Dahlab University of Blida, Soumaa, Algeria

[2]International Institute for Advanced Aerospace Technologies of Saint-Petersburg State University of Aerospace Instrumentation, 67 Bolshaya Morskaya, Saint Petersburg, Russia

ABSTRACT

In this paper, adaptive sensor fusion INS/GNSS is proposed to solve specific problem of non linear time variant state space estimation with measurement outliers, different algorithms are used to solve this specific problem generally occurs in intentional and non-intentional interferences caused by other radio navigation sources, or by the GNSS receiver's deterioration. Non linear approximation techniques such as Extended Kalman filter EKF, Sigma Point Kalman Filters such as UKF and CDKF are computed to estimate the navigation states for UAV flight control. Several comparisons are conduced and analyzed in order to compare the accuracy and the convergence of different approaches usually applied in navigation data fusion purposes. The last non linear filter algorithm developed is the Cubature Kalman Filter CKF which provides more accurate estimation with more stability in Tracking data fusion application. In this work, CKF is compared with SPKF and EKF in ideal conditions and during GNSS outliers supposed to occur during specific interval of time, innovation based adaptive approach is selected and used to modify the covariance calculation of the non linear filters performed in this paper. Interesting results are observed, discussed with real perspectives in navigation data fusion for real time applications. Three parallel modified algorithms are simulated and compared to non-adaptive forms according to Root Mean Square Error (RMSE) criteria.

Keywords: IMU; MEMS; GPS; GNSS; Kalman Filtering; Cubature Rule; Sigma Points; Unscented Kalman Filter

1. INTRODUCTION

Data fusion for non linear system is one of the most important and challenging problems in Multisensor signal processing and integrated navigation systems today. In our work, sensors used are inertial as the main system with external aid provided by GPS and GLONASS receivers known recently as Global Navigation Satellite System "GNSS" solutions. Data fusion based on IMU/GNSS has been widely explored and experimented in the specialized literature [1,2]. For inertial sensors, accelerometers and gyroscopes, the technology of manufacturing these sensors has a great importance and high impact on the accuracy of inertial navigation systems. In this paper, it is assumed that accelerometers and gyroscopes are in the category "Low cost" which gives more important interest in real time applications where most sensors are MEMS Micro Electrical Mechanical Systems based technology. The most inconvenient of these inertial sensors are the biases and drifts growing during time, which needs to be bounded by another technology of sensors such as used in our work, called GNSS receivers. Satellite-based systems such as GNSS gives today's satellite trajectory and high-precision navigation. Inertial sensors combined with GNSS receiver are a good alternative and reliable integrated system for navigation purposes. However, "GNSS"; Galileo E5a/E5b signals and the GPS L5 signal lie within the Aeronautical Radionavigation Services (ARNS) band. They suffer interference from the services in this frequency band, in particular, high power pulsed signals from Distance Measuring Equipment (DME) and Tactical Air Navigation (TACAN) systems embeeded on most aircrafts. The pulsed interference degrades received Signal to Interference and Noise Ratio (SINR), lowers the acquisition sensitivity and even causes the tracking loops to diverge. To maintain system accuracy and integrity, interference mitigation is beneficial and necessary. Adaptive integrated navigation systems INS/GNSS is then proposed for different aerospace applications. However in our work, we focus on GNSS outliers caused by multipath scenario, a bad satellite visibility due to flights in canion environment, or due to non-intentional interferences caused by multiple GSM signals, multiple satellite communication technologies such as Iridium, Globalstar etc. Nevertheless, the classic form of INS/ GNSS data fusion is not adaptive against jamming, impulsive noise, missing measurement... etc. In this paper a solution is proposed based on adaptive extended Kalman filter (EKF) then, compared with more advanced and modern approaches in non linear filtering such as the adaptive SPKF and the adaptive CKF [3].

2. INERTIAL SENSORS

Inertial Measurement Unit "IMU" is the kernel of any inertial navigation system. It is composed by 03 accelerometers and 03 gyroscopes in addition to 03 magnetometers in most modern IMU's. The technology used during the last 50 years has been divided in two principal technologies: Gimbaled INS and Strapdown INS. In our work, the model used is related to strapdown technology with fixed inertial sensors MEMS based, in parallel with body axes. Most of todays inertial sensors are micro electromechanical systems (MEMS). This technology was first used for commercial purposes in the 1990's, and enabled new applications through high miniaturization and cost reduction. Inertial sensors began to be used in completely new domains, such as Pedestrian navigation. However, this miniaturization and cost reduction infuences the performance of the accelerometers and gyroscopes, which explains why some inertial sensors based on previous technologies are still used for high-performance purposes. Gimbaled INS are mechanical with special Horizontal stabilization control algorithm with very expensive cost, they are usually used onboard satellites, spacecrafts, submarine etc.

IMUs based on MEMS sensors are strap-down systems, which means the sensor's orientation depends of the orientation of the object it is on. Theoretically, all types of previously shown MEMS inertial sensors can be combined in an IMU.

2.1. Mechanization of Inertial Measurement Unit

Inertial navigation system is divided in two principal parts: IMU and Digital Signal Processing of sensors data fusion.

Strapdown INS mechanization is described such as in Figures 1 and 2 with a general diagram of SINS as described, based on inertial sensors output; accelerometers and gyroscopes. This navigation system can't ensure long term accuracy of its output, and depends on external aid such as GNSS in most of aerospace applications [4].

2.2. IMU Sensors Output Integration

Inertial measurement units (IMUs) typically contain three orthogonal rate-gyroscopes and three orthogonal accelerometers, measuring angular velocity and linear acceleration respectively. Ideally, the output of the rate-gyroscopes is written as

$$\omega_b(t) = \begin{bmatrix} \omega_{bx}(t) & \omega_{by}(t) & \omega_{bz}(t) \end{bmatrix}^{\mathrm{T}} \tag{1}$$

In practice, however, the outputs contain errors and are written as the formula given below:

$$\tilde{\omega}_b(t) = \omega_b(t) - \delta\omega_b(t), \tag{2}$$

$$\delta\omega_b(t) = \begin{bmatrix} \delta\omega_{bx}(t) & \delta\omega_{by}(t) & \delta\omega_{bz}(t) \end{bmatrix}^{\mathrm{T}} \tag{3}$$

Integrating this yields the updated attitude information for the system provides the following equation:

$$\frac{\mathrm{d}}{\mathrm{d}t} C_b^n(t) = C_b^n(t)\Omega(t), \tag{4}$$

$$\Omega(t) = \begin{bmatrix} 0 & -\tilde{\omega}_{bz}(t) & \tilde{\omega}_{by}(t) \\ -\tilde{\omega}_{bz}(t) & 0 & -\tilde{\omega}_{bz}(t) \\ \tilde{\omega}_{by}(t) & -\tilde{\omega}_{bz}(t) & 0 \end{bmatrix} \tag{5}$$

$$
\begin{aligned}
&C_b^n\left(t+\Delta t\right) \\
&= C_b^n\left(t\right)\mathrm{e}^{\int_t^{t+\Delta t}\Omega(t)\mathrm{d}t} \\
&= C_b^n\left(t\right)\left(I-\frac{\sin\left(\left|\tilde{\omega}_b\Delta t\right|\right)}{\left|\tilde{\omega}_b\Delta t\right|}\Omega\left(t\right)\Delta t\right. \\
&\qquad\left.+\frac{1-\cos\left(\left|\tilde{\omega}_b\Delta t\right|\right)}{\left|\tilde{\omega}_b\Delta t\right|}\left(\Omega\left(t\right)\Delta t\right)^2\right)
\end{aligned}
\tag{6}
$$

Similarly, accelerometers outputs can be written as

$$
a_b\left(t\right)=\left[a_{bx}\left(t\right)\quad a_{by}\left(t\right)\quad a_{bz}\left(t\right)\right]^{\mathrm{T}}, \tag{7}
$$

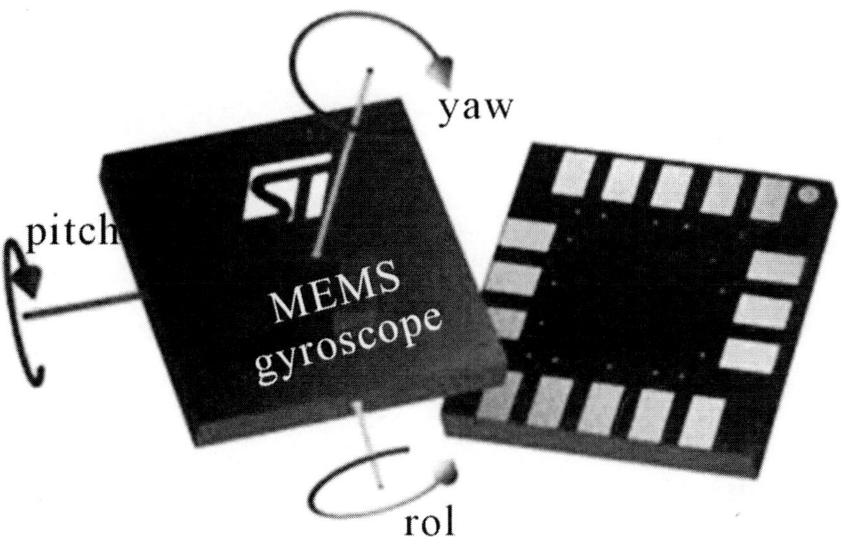

Figure 1. Strapdown Inertial Measurement Unit with MEMS gyroscopes from ST Microelectronic.

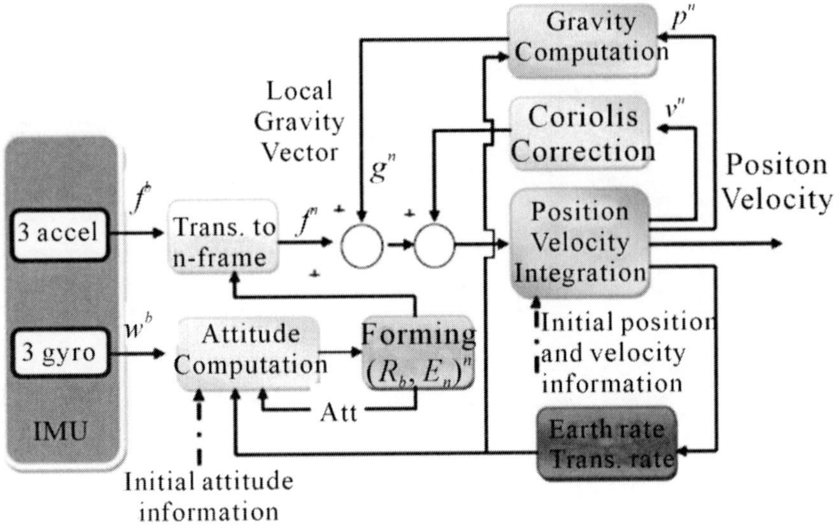

Figure 2. IMU/INS Sensors output integration.

$$\tilde{a}_b(t) = a_b(t) - \delta a_b(t), \tag{8}$$

Two integrations subsequently yield velocity and position updates as follows Velocity integration:

$$V_{n,k} = V_{n,k-1} + \Delta t\left(\tilde{a}_{n,k} - g_n\right) \tag{9}$$

Position integration:

$$Pos_{n,k} = Pos_{n,k-1} + \Delta t\left(V_{ng,k}\right) \tag{10}$$

where g is the estimated gravity vector and Δt is the data period. Collectively, Equations (1) to (10) describe the system model.

2.3. GNSS Global Navigation Satellite System

GNSS signal processing is much explored based on different algorithms tested more and more in real time conditions and in simulations through the specialized literature. We focus on the effect of more satellite visibility in order improve geometry dilution of precision due to the high number of satellites GPS + GLONASS (36 - 40). GLONASS satellites also broadcast signals in the L1 and L2 sub-bands of the radio frequency spectrum as described in **Figure 3**. It is observed in some situation several interferences from different sources for GPS and GLONASS during static and dynamic positioning. GNSS outages or outliers cause accuracy degradation, and sometimes undelivered GNSS receiver positioning.

However unlike PS, GLONASS (Russian) uses frequency division multiple access (FDMA) in both L1and L2 frequency sub-bands. This means that each satellite modulates the same ranging code on carrier signals with slightly different frequencies and is identified by a slot number rather than a Pseudo random Noise (PRN) number. GNSS based on GPS and GLONASS (European system Galileo and Chinese system Compass in the future), are well known satellite navigation ystems and uses parallel positioning techniques; the only difference is that GPS sends different messages on the same frequency (L1, L2, L5) and GLONASS sends the same message on multiple frequencies (L1, L2…). It is important to consider in the near future the new statement of GNSS constellation including Galileo future European system and COMPASS the future Chinese system. Each space constellation has slightly different orbital plane parameters. In this paper, GPS and GLONASS C/A codes are considered in INS/GNSS data fusion. This study concerns non-intentional interferences and outliers/outages in GNSS signal for civilian GNSS receivers. All adjacent communication systems to GNSS band which is a potential source of interferences and have been studied in the literature. In this work direct data fusion technique is applied to an important case when measurement outliers occur affecting GPS and GLONASS receivers during UAV navigation.

2.4. UAV GNC System (Guidance Navigation & Control)

We now describe the application of the SPKF/CKF to the problem of loosely coupled GPS/INS integration for guidance, navigation and control (GNC) of an unmanned aerial vehicle (UAV). The main subcomponents of such a GNC system is a vehicle control system and a guidance & navigation system (GNS) as shown in **Figure 4**. Fixed Wing and Rotor Wing UAVs have been simulated through different dynamical parameters. The embedded system includes an inertial MEMS based IMU, a 10 Hz GPS/ GLONASS synchronized receiver, and a DSP Design Autopilot REVO board for real time implementation. UAV nonlinear control system is based on non linear adaptive state estimators selected as EKF, UKF, CDKF and CKF.

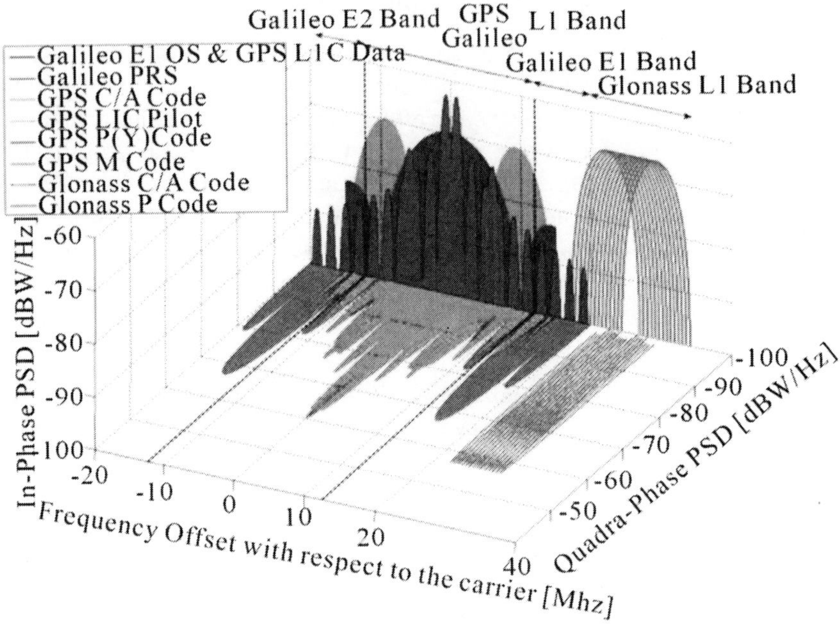

Figure 3. Power spectral densities of GNSS signals.

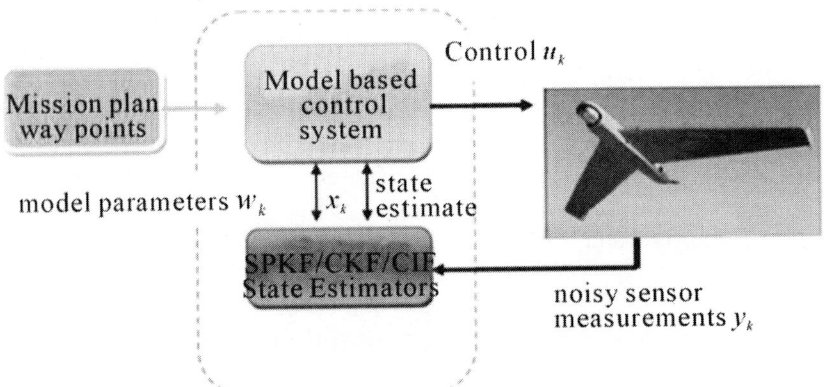

Figure 4. UAV guidance, navigation and control.

We implemented EKF, SPKF (UKF, CDKF) and CKFbased sensor fusion algorithms in normal conditions and during outliers in order to observe the effect on the accuracy and the convergence of filters. We will next discuss the UAV specific system process and observation (measurement) models used inside EKF, SPKF and CKF based navigation system, which is used a second time during GNSS outliers with modified covariance estimation.

We then focus on the application of the Innovation based adaptive CKF/SPKF to the integrated navigation problem. We specifically detail the development of an adaptive based Cubature Kalman Filter ACKF for loosely coupled implementation for integrating GPS measurements with an IMU within the context of autonomous UAV. The next section describe in detail non linear filtring algorithms used and implemented in this work.

3. KALMAN FILTER AND ITS EXTENDED VERSION

In estimation theory, it is well known that for linear state space estimation, affected by white Gaussian noises, the optimal filter is called Kalman filter which is also equivalent to Maximum Likelihood estimator. However in most practical navigation applications, a nominal trajectory does not exist beforehand. The solution is to use the current estimated state from the filter at each time step k as the linearization reference from which the estimation procedure can proceed. Such algorithm is known as extended Kalman filter (EKF).

Extended Kalman Filter EKF

Let us describe below the algorithm of EKF: based on state space model as given by:

$$x_{k+1} = f\left(x_k, u_k, v_k\right) \tag{11}$$

$$y_k = g\left(x_k, w_k\right) \tag{12}$$

Linearization using Taylor approximation at the first order gives the state space model given in most referenced literature. $F_k(.)$is the Jacobian matrix of $f_k(.)$ and $H_k(.)$ is the Jacobian matrix of $h_k(.)$. Thus, following algorithm is obtained:

Initialization:

$$\hat{x}_0 \text{ et } P_0. \tag{13}$$

Prediction:

$$P_{k/k-1} = F_k(\hat{x}_k) P_{k-1} F_k(\hat{x}_k)^T + Q_k \tag{14}$$

$$\hat{x}_{k+1/k} = f_k(\hat{x}_k) \tag{15}$$

Update:

$$K_k = P_{k/k-1} H_k^T(\hat{x}_{k/k-1}) \left[H_k(\hat{x}_{k/k-1}) P_{k/k-1} H_k^T(\hat{x}_{k/k-1}) + R_k \right]^{-1} \tag{16}$$

$$\hat{x}_k = \hat{x}_{k/k-1} + K_k \left[Z_k - h_k(\hat{x}_{k/k-1}) \right] \tag{17}$$

$$P_k = P_{k/k-1} - K_k H_k(\hat{x}_{k/k-1}) P_{k/k-1} \tag{18}$$

The meaning of the extended Kalman filter can be understudied by appreciating the same equation of gain calculation as in the Kalman filter at the difference that in the non linear filtering, EKF is sub-optimal filter. It requires then, more adaptive approaches in solving both filtering and control problems in Aerospace [4,5]. There is another version of extended Kalman filter which could be developed at second order of Taylor approximation, this filter offers better results under high non linearity of the system's dynamic and measurement model. In the next section, another approach based on cubature rule technique is developed.

4. SIGMA POINT KALMAN FILTERING (CUBATURE RULE)

Different estimators were introduced to solve non linear estimation problems; Sigma points Kalman filters (SPKF) introduced in the recent advances in nonlinear filtering. Both Unscented filters (UKF) and (CDKF) mean SPKF, in this case, the density of probability using a deterministic sigma points is estimated at the first and the second order moments of the RGV. For Central Difference Filter, it adopts an alternative method called central difference approximation. Like UKF, CDKF generates several points about the mean based on varying the covariance matrix along each dimension. It evaluates a non linear function at two different points for each dimension of the state vector that are divided by an appropriate chosen interval, SPKF are strong estimators and superior alternative to the EKF in several applications with high non linearity.

4.1. Cubature Kalman Filter CKF

CKF is known as the closest known approximation to the Bayesian filter for non linear estimation with Gaussian assumptions. Such as for UKF and CDKF, CKF doesn't require any Jacobian matrix computation of linearization. The basic steps are given in the next paragraphs. The key assumption of CKF is that the predictive density $p(x_k/Z_{k-1})$ where Z_{k-1} denotes the history of input measurement pairs up to k − 1, and the filter likelihood $p(z_k/Z_k)$ are both Gaussian, which leads to a posterior Gaussian density $p(x_k/Z_k)$. Under these conditions, CKF reduces the computation of mean and covariance with more accuracy. The cubature based Gaussian filter algorithms use cubature rules of the form:

$$I(f) \approx \sum_{i=1}^{m} \omega_i f(\xi_i) \tag{19}$$

to approximate the integral of the form:

$$\int g(x) \mathrm{N}_{\Sigma}^{\mu}(x) \mathrm{d}x = \frac{1}{\sqrt{\pi^n}} \int g\left(\sqrt{2\Sigma}x + \mu\right) \mathrm{e}^{-x^T x} \mathrm{d}x \tag{20}$$

Equation (26) is an integral of a non linear function multiplied by Gaussian weight. The unscented transformation can also be interpreted as an approximation of the integral of the form (27). The technique introduced is based on Gaussian sum filters explored and given in detail by [6]. However it is proposed to model jamming GNSS signal by particular kind of Gaussian sum noise which is twin Gaussian sum affecting only measurement equation. Bellow the algorithm of Adaptive CKF proposed and applied in this work:

4.1.1. Prediction Step

1) Draw cubature points ξ_i, $i = 1, 2, \cdots, 2n$ from the intersections of the n-dimensional unit sphere and the Cartesian axes. Scaled by $\sqrt{n}$. We can write then:

$$\xi_i = \begin{cases} \sqrt{n}e_i \\ -\sqrt{n}e_{i-n} \end{cases} \text{for } i = 1, \cdots, n, i = n+1, \cdots, 2n1 \tag{21}$$

2) Propagate cubature points. The matrix square root is the lower triangular Cholesky factor.

$$X_{i,k-1/k-1} = \sqrt{P_{k-1/k-1}}\xi_i + m_{k-1/k-1} \tag{22}$$

3) Evaluate the cubature points with dynamic model function:

$$\chi^*_{i,k/k-1} = f\left(X_{i,k-1/k-1}\right) \tag{23}$$

4) Estimate the predicted state mean:

$$m_{k/k-1} = \frac{1}{2n}\sum_{1}^{2n} \chi^*_{i,k/k-1} \tag{24}$$

5) Estimate the predicted error covariance:

$$P^-_{x_k} = \frac{1}{2n}\sum \chi^*_{i,k/k-1}\chi^T_{i,k/k-1} - m_{k/k-1}m^T_{k/k-1} + Q^2 \tag{25}$$

4.1.2. Update Step

1) Draw cubature points ξ_i, $i = 1, 2, \cdots, 2n$ from the intersections of the n-dimensional unit sphere and the Cartesian axes. Scaled by $\sqrt{n}$ (as in step 1).

2) Propagate the cubature points.

$$X_{i,k/k-1} = \sqrt{P_{k/k-1}}\xi_i + m_{k/k-1} \tag{26}$$

3) Evaluate the cubature points with the measurement model.

$$Y_{i,k/k-1} = h\left(X_{i,k/k-1}\right) \tag{27}$$

4) Estimate the predicted measurement:

$$\hat{y}_{k/k-1} = \frac{1}{2n}\sum_{i=1}^{2n} Y_{i,k/k-1} \tag{28}$$

5) Estimate the innovation covariance matrix.

$$S_{k/k-1} = \frac{1}{2n}\sum_{i=1}^{2n} Y_{i,k/k-1} Y_{i,k/k-1}^{T} - \hat{y}_{k/k-1}\hat{y}_{k/k-1}^{T} + R_k \tag{29}$$

6) Estimate the cross covariance matrix.

$$P_{x_k y_k/k-1} = \frac{1}{2n}\sum_{i=1}^{2n} X_{i,k/k-1} Y_{i,k/k-1}^{T} - m_{k/k-1}\hat{y}_{k/k-1}^{T} \tag{30}$$

7) Estimate the Kalman gain.

$$K_k = P_{xy,k/k-1} S_{k/k-1}^{-1} \tag{31}$$

8) Estimate the update state.

$$m_k = m_{k/k-1} + K_k\left(y_k - \hat{y}_{k/k-1}\right) \tag{32}$$

9) Estimate the error covariance

$$P_{k/k} = P_{k/k-1} - K_k S_{k/k-1} K_k^{T} \tag{33}$$

NOTE: Comparing with SPKF, there are no parameters to tune in CKF approximating non linear functions of the system and measurement. Another alternative to approximate the lower bound for non linear state estimation with additive Gaussian noises is given and described in [7,8].

NOTE: Comparing with SPKF, there are no parameters to tune in CKF approximating non linear functions of the system and measurement. Based on similar idea such as for sub optimal fading factor, it is possible to combine Sigma Point Kalman Filters (UKF, CDKF, CKF) and adaptive fading approach.

4.2. Adaptive Cubature Kalman Filter ACKF

It is possible to define then the fading factor such as given bellow:

$$\lambda_{i,k} = \frac{\mathrm{tr}\left[\eta V_k - \varepsilon R_k\right]}{\mathrm{tr}\left[P_{yy}\right]} = \begin{cases} \lambda_{i,k}, & \rightarrow \quad \lambda_{i,k} > 1 \\ 1, & \quad\quad \lambda_{i,k} \leq 1 \end{cases} \tag{34}$$

With parameters defined below:

$$V_k = \begin{cases} v_0 v_0^{T} \\ \dfrac{\left[\rho V_{k-1} + v_k v_k^{T}\right]}{1+\rho}, k \geq 2 \end{cases} \tag{35}$$

Thus, the covariance matrices need to be updated based on the adaptive fading factor such as given in the following equations:

$$P_{x_k}^{-} = \frac{\lambda_k}{2n} \sum \chi_{i,k/k-1}^{*} \chi_{i,k/k-1}^{T} - m_{k/k-1} m_{k/k-1}^{T} + Q_k \tag{36}$$

$$S_{k/k-1} = \frac{\lambda_k}{2n} \sum_{i=1}^{2n} Y_{i,k/k-1} Y_{i,k/k-1}^{T} - \hat{y}_{k/k-1} \hat{y}_{k/k-1}^{T} + R_k \tag{37}$$

$$P_{x_k, y_k/k-1} = \frac{\lambda_k}{2n} \sum_{i=1}^{2n} X_{i,k/k-1} Y_{i,k/k-1}^{T} - m_{k/k-1} \hat{y}_{k/k-1}^{T} \tag{38}$$

The key parameter in this adaptive algorithm is the fading factor. It requires the defined parameters, some other techniques in literature use multiple fading factors which is not always superior to the single fading factor and are commonly selected.

Conceptually, the implementation principle of SPKF/ CKF resembles that of the EKF, the implementation, however, is significantly simpler because it is not necessary to formulate the Jacobian and/or Hessian matrices of partial derivatives of the nonlinear dynamic and measurement equations, which is very important for real time implementation.

Before, implementing adaptive SPKF and CKF, we propose to observe the effect of Innovation based adaptive EKF on the navigation state during outliers as a first experience, thus, go in more advanced signal processing for UAV data fusion during GNSS outliers.

The 1st Simulation of EKF-Adaptive EKF

The condition of simulation is described in **Table 1**:

T = 50 s, N = 5000; dt = 0.001; g = 9.81 m/s/s; Outlier duration of time ODT = 19 s.

Initial Motion

Distance N, E, D (meter)	Velocity N, E, D (m/s)	Attitude angles pitch, roll, yaw (rad)
1000	260	$10\pi/180$
1000	70	$45\pi/180$
1000	50	$10\pi/180$

From **Figure 5** to **Figure 12**, one can observe the state estimation results of the non linear part described in the previous section, so the north velocity with corresponding MSE, East velocity with corresponding MSE, vertical velocity with its corresponding MSE and pitch attitude observation based on multiple GNSS antenna with the corresponding MSE. The figures content show how accuracy degradation of EKF can be caused by GNSS measurement outliers. In the first part of navigation during reliable GNSS outputs, EKF and Adaptive EKF seem to be superposed.

IMU outputs without observation diverge from the real trajectories and EKF during outliers is less accurate than IMU integration outputs.

Table 1. Initial values of covariance matrices P, Q, R.

Estimation Covariance $\sqrt{diqg(p)}$	System Covariance $\sqrt{diqg(Q)}$	Measurement Covariance $\sqrt{diqg(R)}$
100	10	10
100	10	10
100	10	15
10	2	2
10	2	2
10	2	2
$10\pi/180$	$0.5\pi/180$	$1.5\pi/180$
$10\pi/180$	$0.5\pi/180$	$1.5\pi/180$
$10\pi/180$	$0.5\pi/180$	$1.5\pi/180$

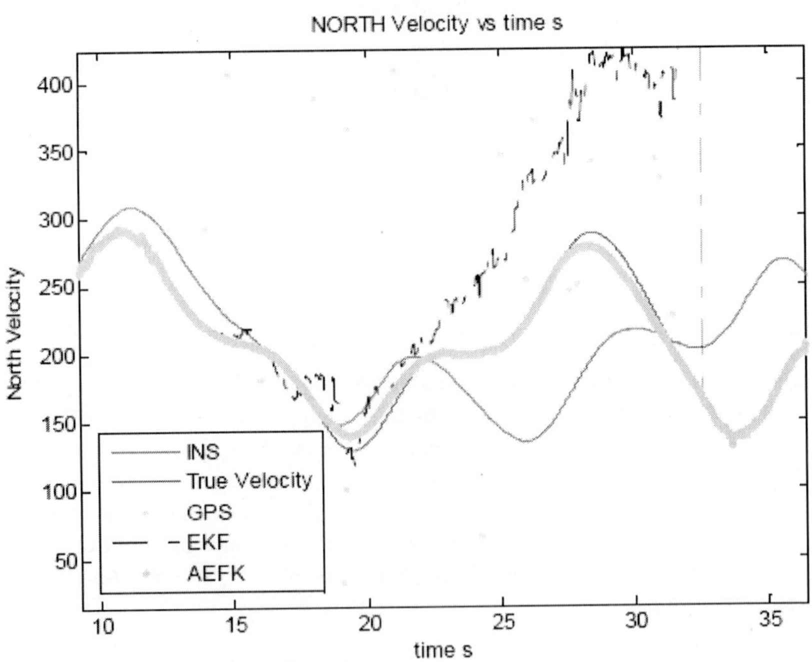

Figure 5. North velocity estimation.

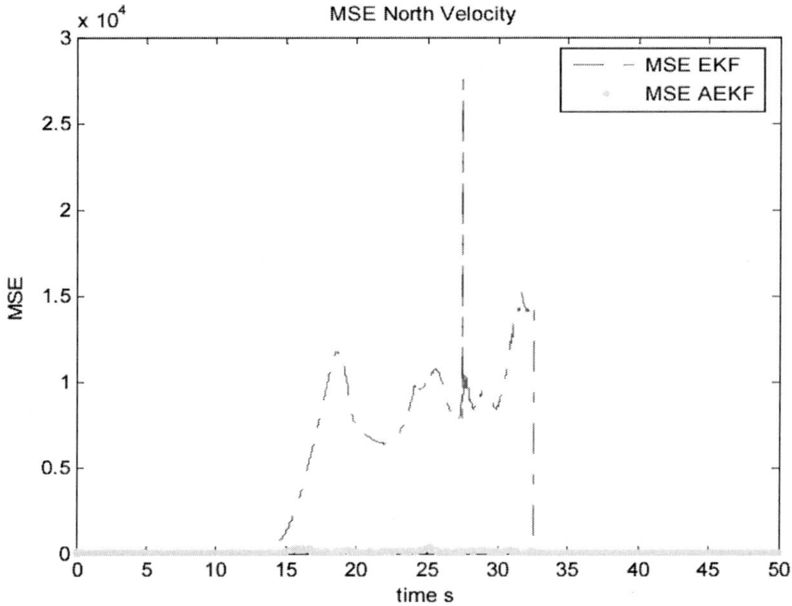

Figure 6. MSE north velocity.

Attitude estimation based on multiple GNSS antennas.

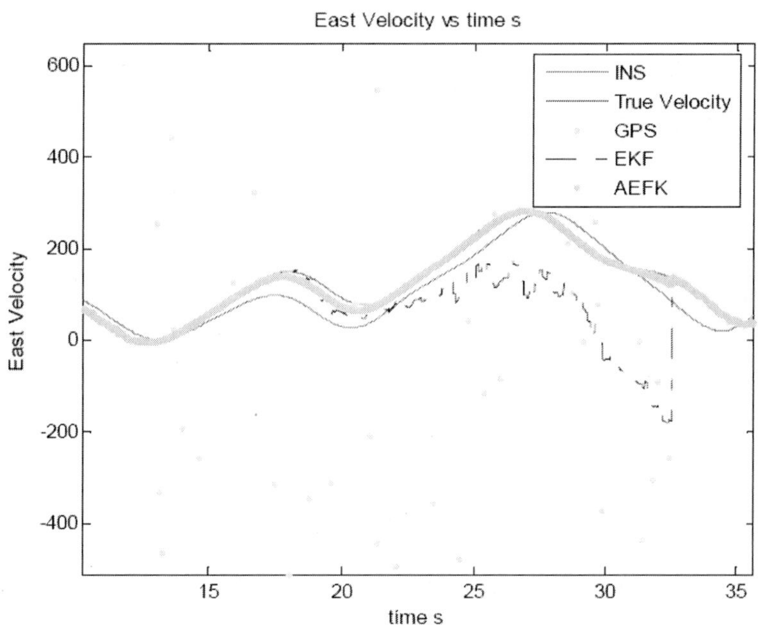

Figure 7. East velocity estimation.

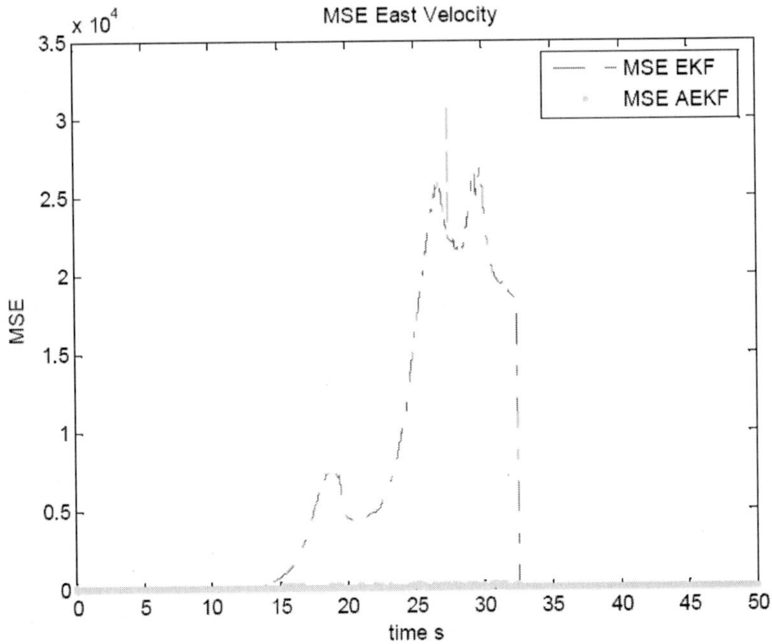

Figure 8. MSE north velocity.

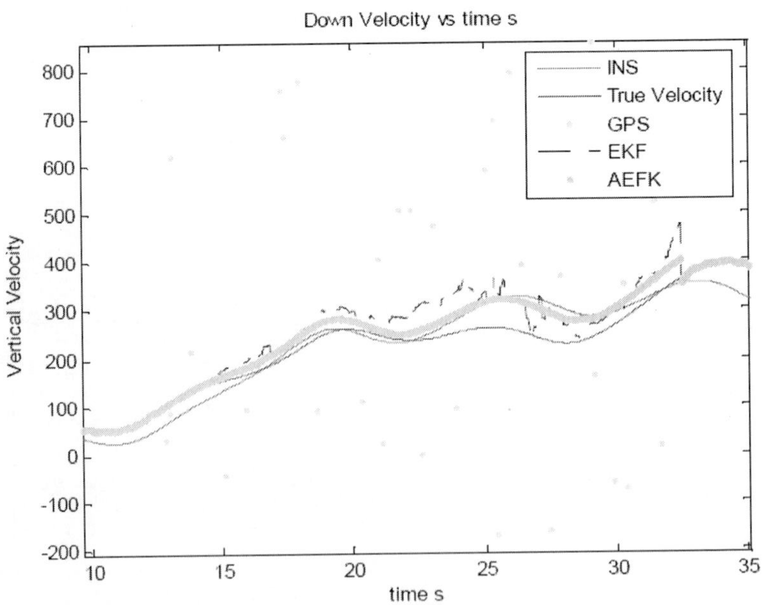

Figure 9. Vertical velocity estimation.

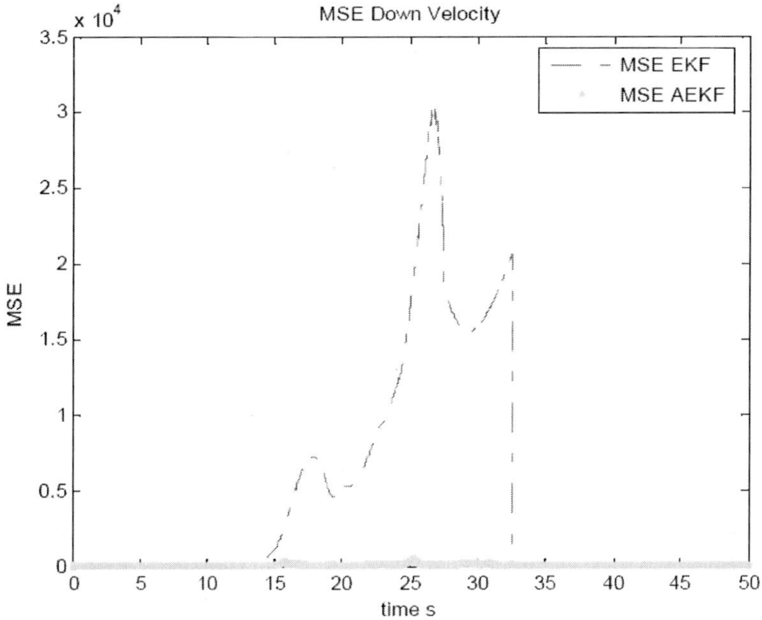

Figure 10. MSE vertical velocity estimation.

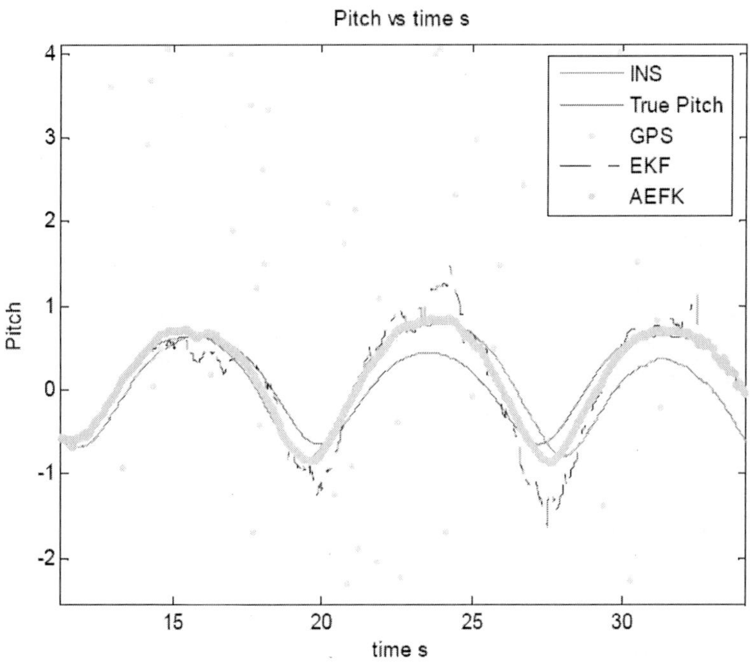

Figure 11. Pitch angle estimation.

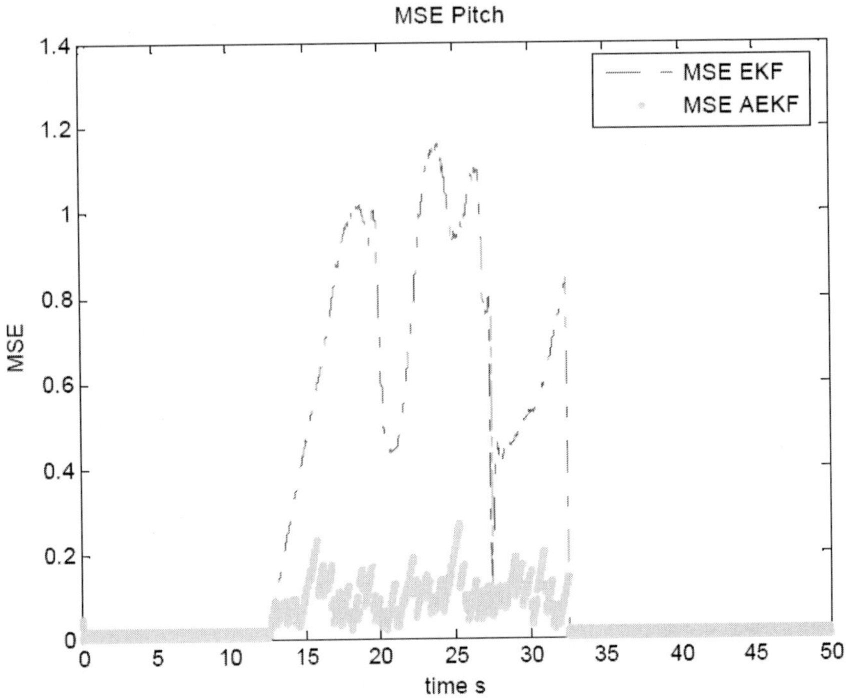

Figure 12. MSE pitch angle estimation.

After multiple simulations, it is clear through state and related MSE estimation that Innovation based Adaptive EKF "AEKF" provide an ameliorated filtering accuracy comparing with EKF. This is confirmed by different state estimation with different non linearity such as velocity and attitude states.

5. SIMULATION OF ADAPTIVE CUBATURE KALMAN FILTER

Let us go back to the system given by (1) and by the measurement noise affected by outliers. Let the initial state be described as a percentage of the true initial state. These next simulations should provide two important observations:

1) Compare EKF,UKF,CDKF and CKF during outliers without modification of Covariances estimation.

2) Compare Innovation based Adaptive EKF, UKF, CDKF and CKF during GNSS outliers after modification of covariance estimation algorithm.

5.1. Results and Discussion

The same simulation conditions of the previous section are considered in this second experience: first, attitude estimation is presented with yaw, pitch and roll angles estimation. Then, velocity through 03 axes of navigation (north, east, down) are observed and discussed, with finally, the linear part in state estimation presented in this paper with position state vector estimation.

Attitude estimation under outliers from Figure 13 to Figure 18.

Velocity estimation under outliers from Figure 19 to Figure 25.

Linear estimation part: Position.

The only linear part in this work was the location estimate on with north, east and vertical distance states are observed on Figures 26 and 27.

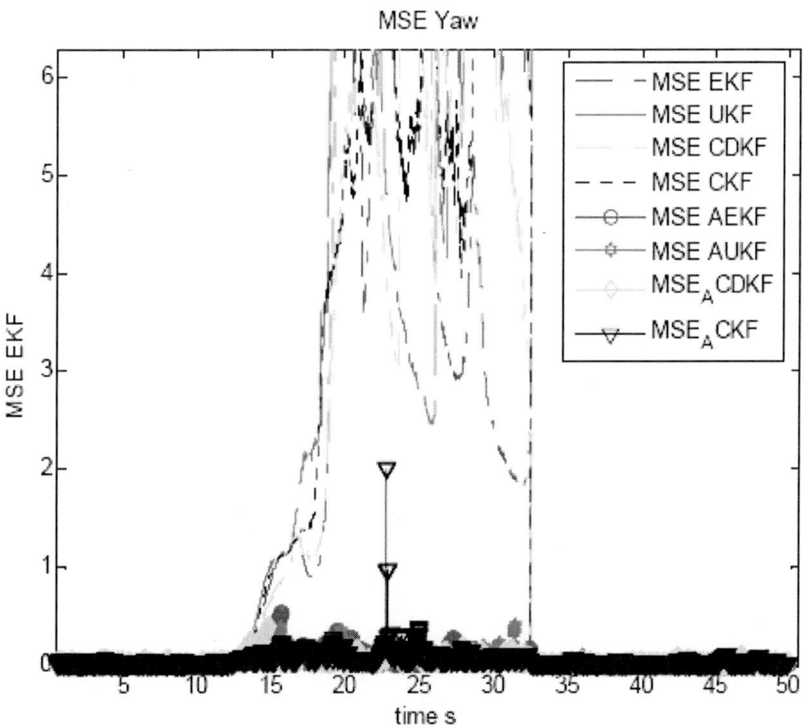

Figure 13. MSE yaw angle estimation in degree.

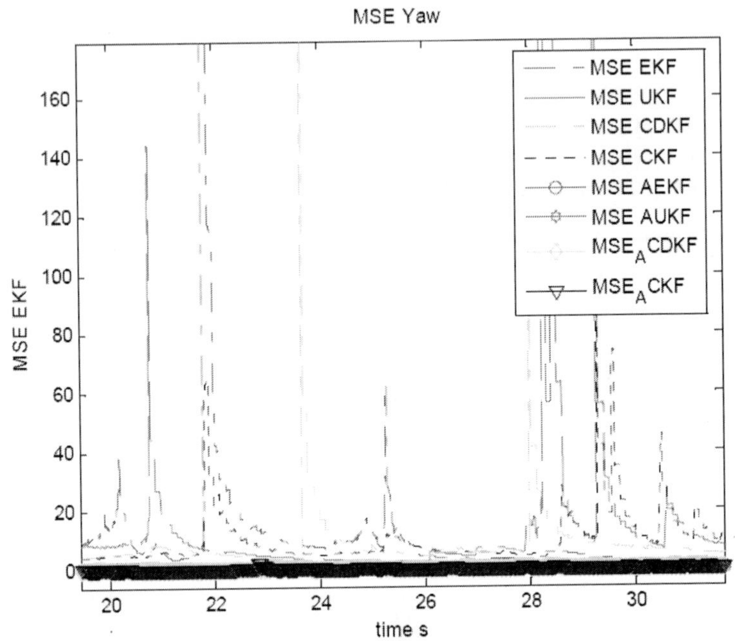

Figure 14. MSE yaw angle estimation in degree zoom.

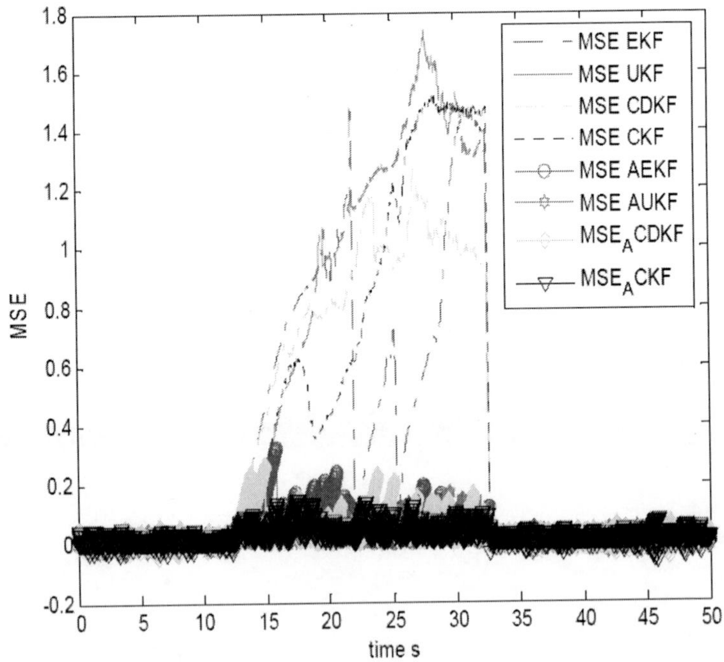

Figure 15. MSE pitch angle estimation in degree.

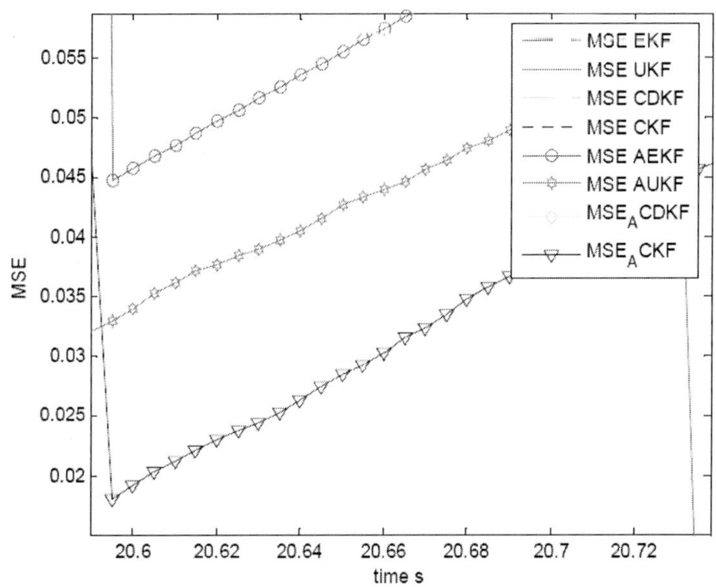

Figure 16. MSE pitch angle estimation in degree zoom.

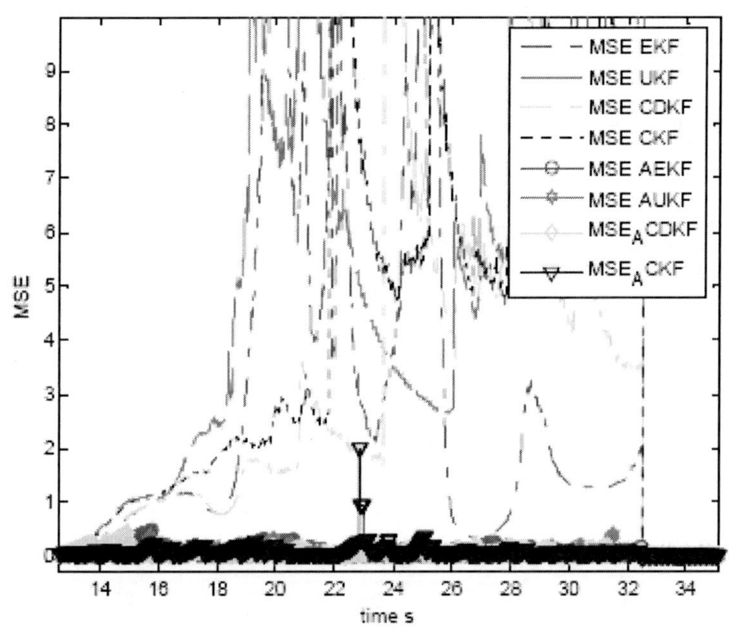

Figure 17. MSE roll angle estimation in degree.

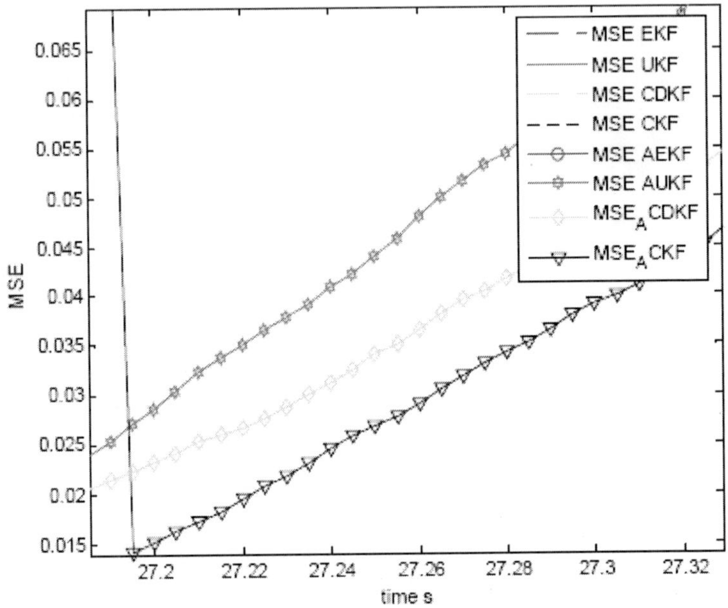

Figure 18. MSE roll angle estimation in degree zoom.

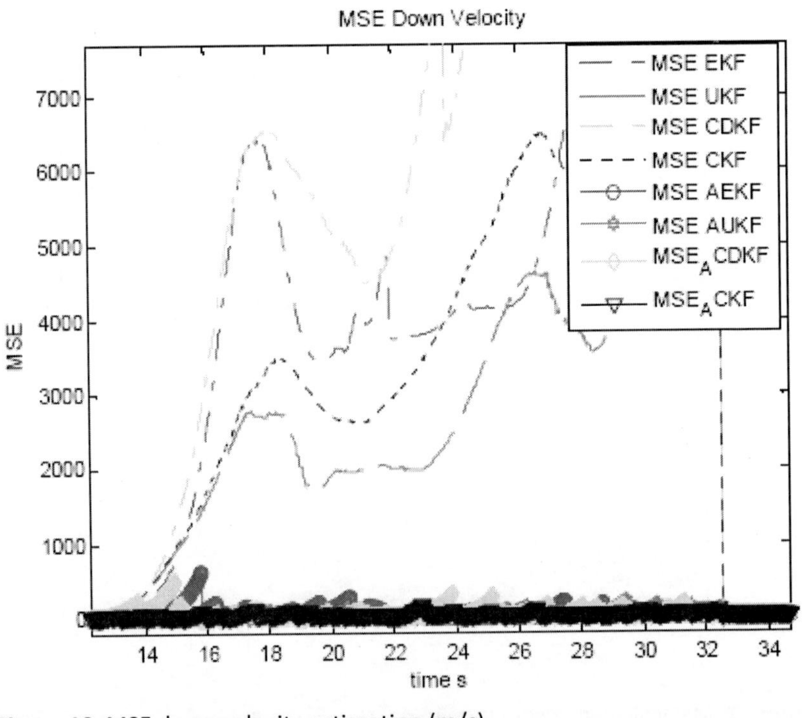

Figure 19. MSE down velocity estimation (m/s).

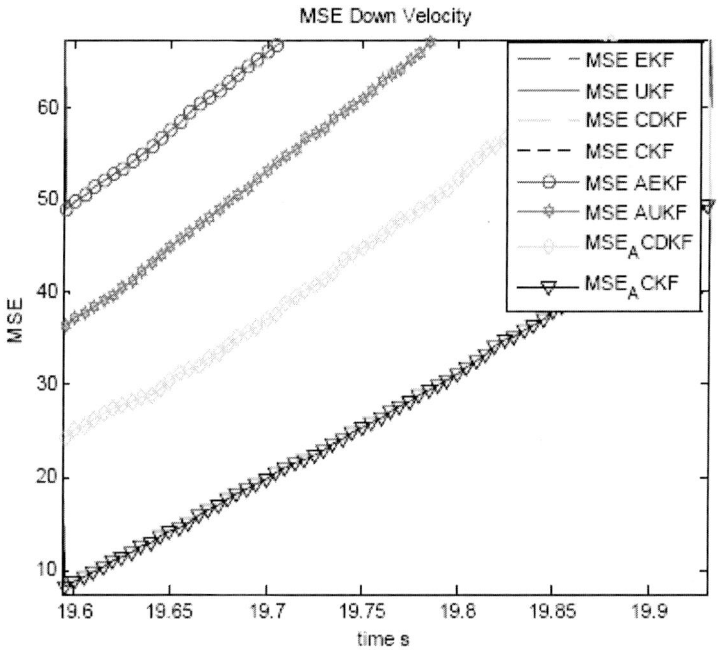

Figure 20. MSE down velocity estimation (m/s) zoom.

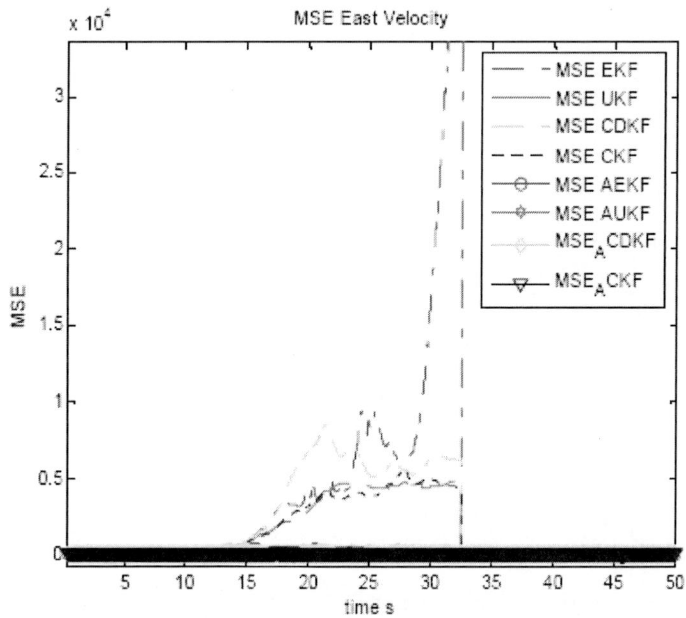

Figure 21. MSE east velocity estimation (m/s).

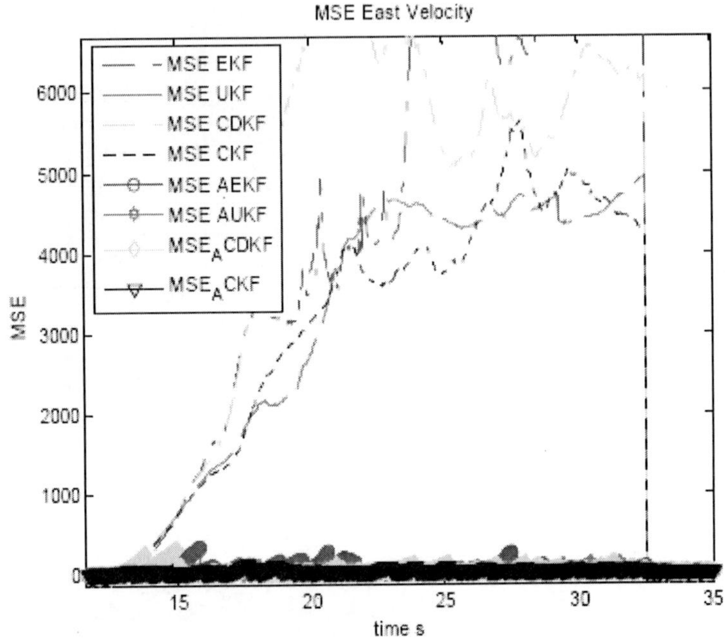

Figure 22. MSE east velocity estimation (m/s).

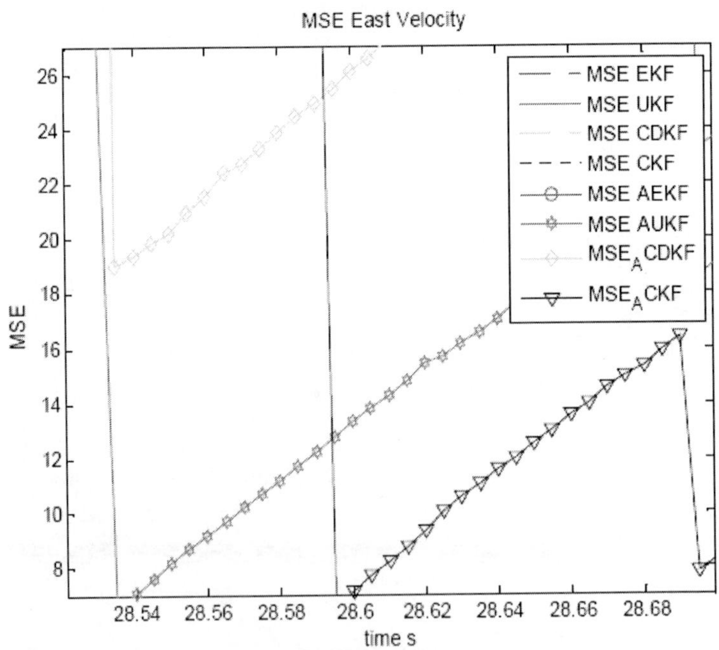

Figure 23. MSE east velocity estimation (m/s) zoom.

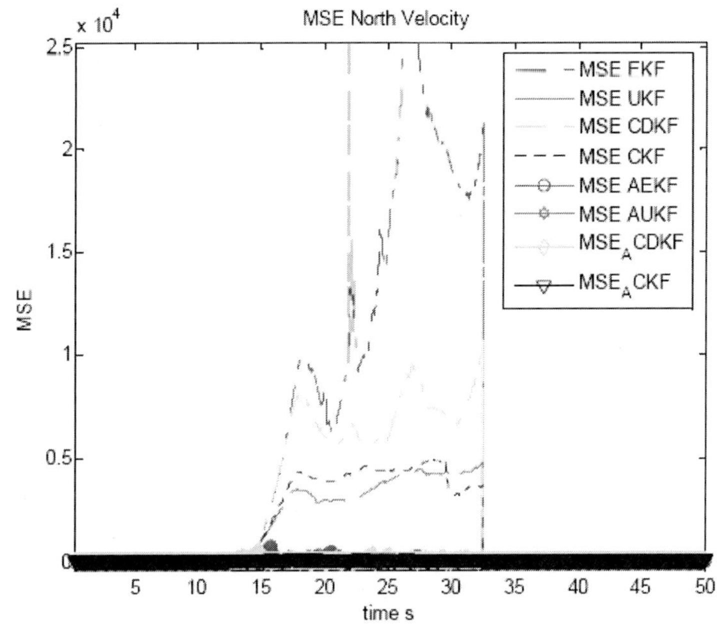

Figure 24. MSE north velocity estimation (m/s).

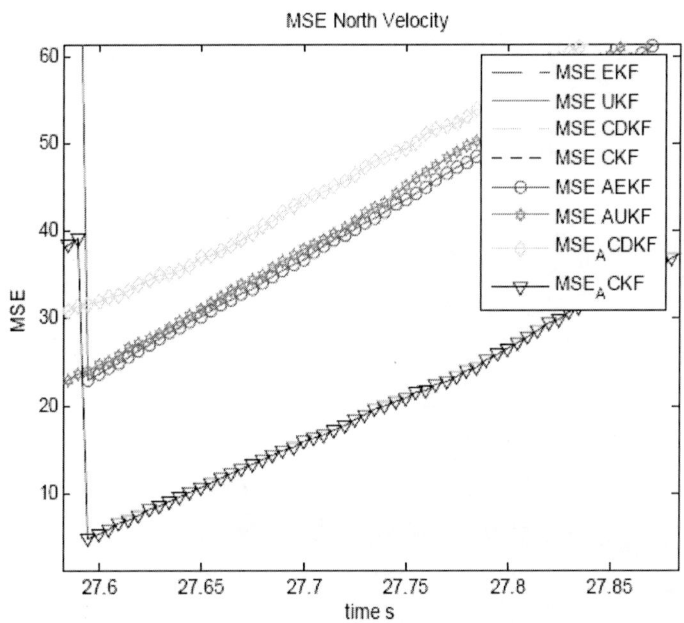

Figure 25. MSE north velocity estimation (m/s) zoom.

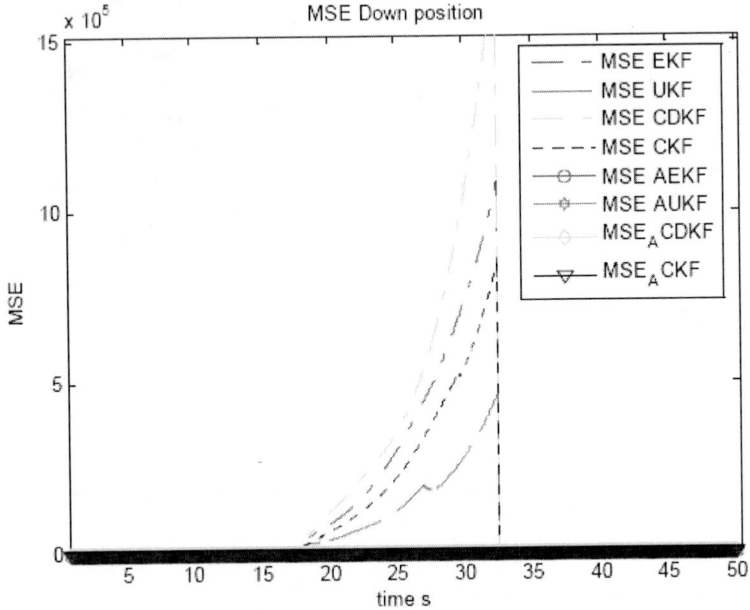

Figure 26. MSE vertical position estimation (m/s).

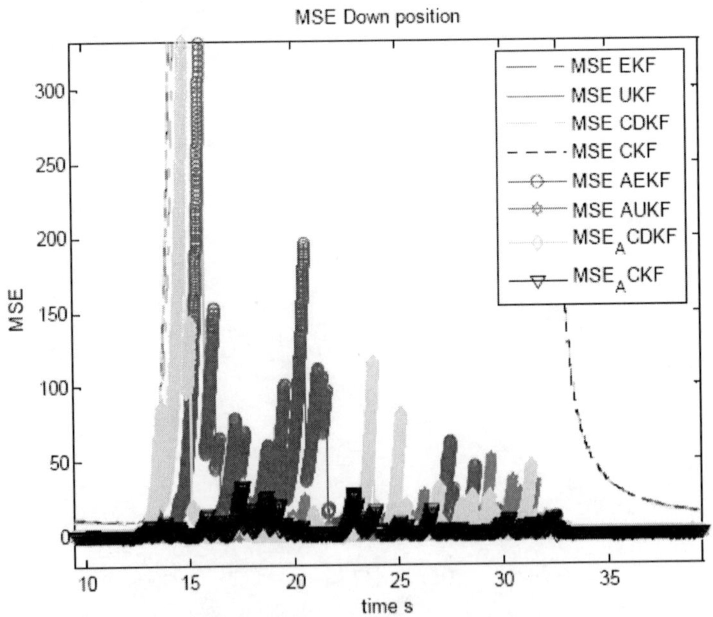

Figure 27. MSE vertical position estimation (m/s) zoom.

5.2. Observation of Adaptive Non Linear State Estimation for IMU/GNSS Data Fusion during Outliers

During simulations, the following algorithms are compared and applied ton MEMS IMU/GNSS data fusion during 50 seconds. In this interval, GNSS outliers are simulated and activated between 13 s and 32 s. For the first part of simulation in normal flight conditions, between 0 and 13 s, it is possible to observe that all filters, classical and adaptive are superposed and present the same accuracy and convergence curve. This is logic results due to unuseful innovation based adaptive fading factors used to improve the accuracy of filtering algorithms during outlier period. It is very interesting to observe the clear part of all figures which describes classical non linear filters EKF, CDKF, CKF and UKF without any adaptation, except for velocity estimation which in that case, UKF performs CKF in its non adaptive forms during outliers. This is an unexpected result between UKF and CKF. In this precise situation, UKF has better accuracy than CKF which is not usual in previous study and experiments in the field of data fusion. This is considered as a interesting novel result. As for EKF and CDKF, it is possible to observe too an additional interesting result, CDKF RMSE is less accurate than EKF RMSE during outlier period for velocity estimation which means that GNSS outliers affect more the CDKF covariance estimation than the EKF computation.

It is then possible to distinguish a classification of filters which is for most states estimation according the accuracy and convergence of filters: EKF, CDKF, UKF, CKF during outliers. It is very interesting to observe in the second curves of the adaptive filters the clear part of all figures which describes adaptive non linear filters AEKF, ACDKF, AUKF and ACKF, ACKF performs all adaptive ASPKF and AEKF during outliers. It is clear that ACKF maintain RMSE close to the minimum variance which is not the case for AUKF, ACDKF and AEKF, those algorithms present some impulsive errors during outliers even with innovation based adaptive fading factor. This is the proof that CKF after modification of its covariance and mean estimation presents a superior accuracy and stability of convergence comparing with modified SPKF and EKF. It is also clear that ACKF performs AUKF, ACDKF and finally AEKF. The last figures of position estimation, it is possible to observe the RMSE of ACKF, ASPKF and AEKF during outliers, and it is clear that for this linear state, non linear filters are non adequate only ACKF provide a good results between 9 s and 32 s. It is then possible to generalize the observation that all transformed filters AEKF, AUKF, ACDKF and ACKF outperform their classical algorithms EKF, UKF, CDKF and CKF respectively. This means that the solution we propose is a good alternative to GNSS outliers in IMU/GNSS data fusion.

6. CONCLUSION

The design of non linear adaptive filters and the associated data fusion based on EKF, SPKF and CKF were deeply studied for MEMS IMU/GNSS sensor fusion during satellite positioning outliers. Based on the innovation fading factor, 04 non linear filtering approaches EKF, KF, CDKF and CKF were modified using this fading factor for covariance and mean estimation during GPS/ LONASS outages. By the way, several solutions were proposed and ameliorated the accuracy and the time of convergence of each modified filtering algorithm. However according the RMSE criteria, Cubature Kalman Filter CKF during outliers demonstrated some limitation to perform Unscented Kalman Filter UKF in its non adaptive form. Finally, adaptive fading CKF demonstrated a large superiority in estimation accuracy and in its stability comparing with other filtering classes—adaptive EKF, adaptive UKF and adaptive CDKF.

REFERENCES

1. I. Arasaratnam and S. Haykin, "Cubature Kalman Filtering," IEEE Transactions on Automatic Control, Vol. 54, No. 6, 2009, pp. 1254-1269.
2. A R. Runnalls, "A Kullback-Leibler Approach to Gaussia Mixture Reduction," IEEE Transaction on Aerospace and Electronic System, 2006.
3. P. Tchikavsky, C. H. Muravchik and A. Nehorai, "Posterior Cramer Rao Bounds for Discrete Time Nonlinear Filtering," IEEE Transactions on Signal Processing, Vol. 46, No. 5, 1998, pp. 1386-1396.
4. H. Benzerrouk and A. V. Nebylov, "Robust Nonlinear Filtering Applied to Integrated Navigation System under Non Gaussian Measurement Noise Effect," Proceeding of IEEE AEROSPACE Conference 2012, Big Sky Montana.
5. D. Karlgaard Christopher and S. Henspeter, "Huber Divided Difference Fiters," Vol. 30, No. 3, 2007, pp. 885- 891.
6. P. Closas and C. Fernandez-Prades, "Bayesian Nonlinear Filters for Direct Position Estimation," Proceedings of the IEEE Aerospace Conference, Big Sky, MT (USA), March 2010, pp. 1-12.
7. P. Closas and C. Fernandez-Prades and J. Vilà-Valls, "Multiple Quadrature Kalman Filtering," IEEE Transactions on Signal Processing, Vol. 60, No. 12, 2012, pp. 6125-6137.

8. H. Benzerrouk, "Gaussian vs. Non-Gaussian Noise in Inertial/GNSS integration," GNSS Solutions, Inside GNSS Magazine, November/December 2012, pp. 32 39.

Chapter 8

An Adaptive Compensation Algorithm for Temperature Drift of Micro-Electro-Mechanical Systems Gyroscopes Using a Strong Tracking Kalman Filter

Yibo Feng, Xisheng Li * and Xiaojuan Zhang

School of Automation and Electrical Engineering, University of Science and Technology Beijing, Beijing 100083, China

ABSTRACT

We present an adaptive algorithm for a system integrated with micro-electro-mechanical systems (MEMS) gyroscopes and a compass to eliminate the influence from the environment, compensate the temperature drift precisely, and improve the accuracy of the MEMS gyroscope. We use a simplified drift model and changing but appropriate model parameters to implement this algorithm. The model of MEMS gyroscope temperature drift is constructed mostly on the basis of the temperature sensitivity of the gyroscope. As the state variables of a strong tracking Kalman filter (STKF), the parameters of the temperature drift model can be calculated to adapt to the environment under the support of the compass. These parameters change intelligently with the environment to maintain the precision of the MEMS gyroscope in the changing temperature. The heading error is less than 0.6° in the static temperature experiment, and also is kept in the range from 5° to −2° in the dynamic outdoor experiment. This demonstrates that the proposed algorithm exhibits strong adaptability to a changing temperature, and performs significantly better than KF and MLR to compensate the temperature drift of a gyroscope and eliminate the influence of temperature variation.

Keywords: Strong tracking Kalman filter; Bias; compass; MEMS gyroscope

1. INTRODUCTION

MEMS inertial sensors are more widely used in the middle- and low-level market because of their advantages such as diminutive size, low cost, and low power consumption [1]. However, the temperature variation reduces the precision of the MEMS gyroscope seriously when the MEMS gyroscope is used in a changing environment such as in the car or unmanned aerial vehicle without a constant temperature adjustment [2], and this feature makes it necessary to model and compensate for the bias drift (called "drift" for short) of the MEMS gyroscope to increase the accuracy.

The noise in an MEMS gyroscope is composed of two components: a slow-changing component and a high-frequency component with an average of zero. The gyroscope is sensitive to temperature variations, so the surrounding temperature variation leads to the bias drift of the gyroscope. Then, as the error of the angular velocity, the drift causes error accumulation in the orientation, thus, this is the major part of the slow-changing component [3]. The drift is not linear with temperature, therefore, the model is not entirely accurate. Additionally, different MEMS gyroscopes may have the same drift model, however, because of the differences in the respective production process, the parameters of each gyroscope vary between one another. Thus, the result is not accurate if we use the same parameters to compensate the drifts of all gyroscopes. However, the cost and time of calibration will increase if we calibrate the parameters of every MEMS gyroscope, and the parameters will be also changed to affect the accuracy of the gyroscope when the supply voltage or work environment changes [4]. Thus, we present an adaptive algorithm for a system integrated with MEMS gyroscopes and a compass to compensate the drift of MEMS gyroscopes for different temperatures and work environments and to improve the accuracy of the MEMS gyroscope.

Before explaining our algorithm, it is necessary to briefly discuss several other approaches for compensating the drift of a MEMS gyroscope. Kirkko-Jaakkola *et al.* demonstrated that temperature variation greatly affects the bias of the MEMS gyroscope [5]. Consequently, temperature drift compensation should be implemented to reduce the influence of temperature. There are two common ways to implement the compensation. The first way is using a temperature control system to keep the temperature stable. Because of the stable temperature, the bias is not influenced by temperature, so this method has high precision and reduces the computational burden, and also makes the use of the temperature-sensitive instruments much easier, but the cost, power consumption and size are all increased [6]. The second way is using signal processing to compensate the temperature drift. This way does not increase the size, weight and power consumption, and it analyses the characteristics of the drifts, then uses a suitable model to compensate the drift, but the precision depends on the accuracy of the model and parameters. As we know, the relation between the drift and the temperature is nonlinear, so if the model and parameters are more accurate, the compensation precision is higher, but the more complex model increases the computational burden.

Temperature drift modeling and compensation is used much more widely than the first way. Firstly, the model of the temperature drift is built by multiple linear regression (MLR) on the basis of the structure of the gyroscope and experimental data. Then, calibration experiments are necessary to obtain the available data. Finally, algorithms such as the least squares method, or polynomial segmentation fitting are used to calculate the parameters of the model. Afterwards, temperature compensation can be implemented [7,8,9,10,11]. These methods are easy to implement and compensate the drift in the full temperature range. However, the main drawback of these approaches is that the relationship between the bias of the gyroscope and the temperature is complicated and nonlinear, so the model is not completely accurate, so the fixed parameters cannot maintain high compensation precision over the full temperature range. Additionally, if the working environment is altered, the accuracy will be affected, because the parameters do not change to adapt to the environment. To reduce the error caused by inaccuracy of the model and improve the fault-tolerance, the artificial neural network is used for the temperature compensation of gyroscope bias, but the features of the neural network, such as low convergence speed and easily falling into local minima, limit the compensation accuracy [12,13]. Some methods, such as Kalman Filter (KF), extended Kalman Filter (EKF), iterated unscented Kalman Filter (IUKF), can be used for neural network training to reduce the influence of these features and improve the compensation precision [14,15]. KF, EKF, UKF are the most common methods for integrated navigation systems to compensate various errors which include gyroscope drift. The compensation precision is high and the estimations of errors change after each iteration to match the environment when each navigation subsystem works properly, so the gyroscope drift is compensated accurately with the support of the other navigation subsystems [16,17,18], but the estimation of gyroscope drift changes only after iteration. Once a fault occurs at some subsystem, the iteration may stop to wait for the fault removal, and the drift estimation also remain unchanged, so faults have a great influence on the compensation precision.

In order to solve the problem that the model is not completely accurate, decrease the influence of faults and accurately compensate the MEMS gyroscope drift, we build a simple drift model and estimate suitable parameters for the current environment after each iteration, so an adaptive compensation algorithm based on STKF is presented in this paper. Firstly, a simplified bias model is built to reduce the amount of calculation. Then the STKF is built on the basis of the simplified bias model, and the bias model parameters are set as state variables of STKF, so these parameters are estimated accurately under the support of a compass, and these parameters make the simplified model maintain high precision in the current environment. When the STKF is stopped for some reason, such as a fault or magnetic interference, the bias drift is also compensated accurately by the model and parameters. This algorithm is different from the other algorithms in that it can estimate the most appropriate parameters for the simplified model in the current environment. If the environment does not change too much, these parameters can maintain a high precision of bias estimation. The parameters change adaptively with the environment, thus, they have a strong adaptability to the changing environment, which makes the compensated gyroscope unaffected by the environment.

The system configuration is introduced in Section 2. Section 3 provides a detailed description of the adaptive compensation algorithm. Section 4 presents the experimental results of this algorithm, and finally, the conclusion of this paper is presented in Section 5.

2. SYSTEM CONFIGURATION

The system is integrated with a triaxial compass, three MEMS gyroscopes (ADXRS623), and a triaxial accelerometer. The triaxial compass is composed of a single-axis magnetoresistive sensor (HMC1021Z) and a dual-axis magnetoresistive sensor (HMC1022), which are used to measure the axial intensity of the magnetic field for every axis. The triaxial accelerometer is composed of a single-axis accelerometer (ADXL103) and a dual-axis accelerometer (ADXL203), which are used to measure the axial intensity of the acceleration for every axis to calculate the attitude angle of the system. Then, the orientation of the compass can be calculated from the axial strength of the magnetic field, which is compensated by the attitude angle of the system [19]. The characteristics of these sensors are shown in Table 1. The orientation is calculated using the arctangent of the axial intensity of the magnetic field which is compensated by the axial intensity of the magnetic field on z-axis and attitude angles, so the error is not accumulated. Additionally, the impacts of temperature variation on the magnetoresistive sensors are at almost the same proportion, so the arctangent can mostly eliminate the impact from temperature variation. The orientation of the compass can be used to assist in the compensation of the MEMS gyroscope bias drift.

Table 1. The characteristics of sensors.

	ADXRS623	HMC1022/1021	ADXL103/203
Measurement range	±150°/s	±6 gauss	±1.7 g
Sensitivity	12.5 mv/°/s	1 mv/v/gauss	1000 mv/g
Sensitivity change due to temperature	±3%	−0.3%/°C	±0.3%
Noise Density	0.04°/s/√hz	48 nv/√hz	110 μg/√hz rms

3. THE ADAPTIVE COMPENSATION ALGORITHM

In this paper, we lump all slow-changing errors together, regardless of whether they are caused by physical phenomena or temperature sensitivity, and call them collectively 'drift'. The system is always used in many types of environments, and the working temperatures are also different, thus, a temperature compensation model is quite necessary, and the errors caused by other error sources can be treated as constants in a short period of time. Therefore, a model of the MEMS gyroscope bias drift can be constructed as the model below, and an STKF is used to estimate the parameters of model.

3.1. The Model of MEMS Gyroscope Drift

The model of the uncompensated angular velocity of the MEMS gyroscope is:

$$\omega = \omega_t + B_d + n \tag{1}$$

where:

ω—measurement of angular velocity.

ωt—true but unknown angular velocity.

Bd—slow-changing component of the signal; this is the gyroscope drift.

n—stochastic component of the signal.

In this paper, the precision of the gyroscope is impacted by Bd and n. We know that Bd, as the slow-changing component, is mainly caused by physical phenomena and temperature variation, so a temperature-compensating model is needed, and errors caused by the other error sources can be treated as constant in a short period of time. The model of Bd can be constructed [20].

The slow-changing component of the gyroscope is not only related to the measured temperature of the MEMS gyroscope but is also related to the temperature gradient of the surroundings. Additionally, the temperature gradient and the rate of temperature variation have a linear relationship; therefore, the slow-changing component of the MEMS gyroscope can be modeled as:

$$B_d = a*T + b*T' + c \tag{2}$$

where:

T—measured temperature of the gyroscope.

T'—rate of temperature variation.

a, b, c—parameters of the model [21].

3.2. The STKF for Model Parameters

The accurate estimated value of the slow-changing component of the MEMS gyroscope requires accurate values of a, b, and c in real time. Consequently, the STKF is modeled on the basis of Equation (2), and a, b, and c are all state variables of the STKF [22]. The STKF is as follows:

$$\begin{aligned} x_{k+1} &= \Phi_{(k+1,k)} * x_k + \Gamma_{(k+1,k)} * \omega_k \\ y_{k+1} &= H_{k+1} * x_{k+1} + v_{k+1} \end{aligned} \tag{3}$$

where:

x_{k+1}—state variables of the STKF at time $k+1$ and also the parameters of the slow-changing component at time $k+1$, $x_{k+1} = \begin{bmatrix} a \\ b \\ c \end{bmatrix}_{k+1}$.

y_{k+1}—measurement of the STKF at time $k+1$, $y_{k+1} = \begin{bmatrix} m_{k+1} \\ m_k \\ m_{k-1} \end{bmatrix}$. As the initial values of y_{k+1}, y_1, y_2 and y_3 is measured when the gyroscope is motionless in the beginning.

m_{k+1}— measurement of the gyroscope drift at time k + 1.

$$m_{k+1} = \hat{z}_k + \Delta A_{(k+1,k)} / \Delta t \quad (4)$$

$\hat{z}_k$— estimated value of the gyroscope drift based on the parameters at time k.

$$\hat{z}_k = a_k * T_k + b_k * T'_k + c_k \quad (5)$$

$\Delta A_{(k+1,k)}$— variation of the difference between the orientation of the compass and the orientation of the MEMS gyroscope from time k to time k + 1.

$$\Delta A_{(k+1,k)} = (A\_compass_{(k+1)} - A\_Gyro_{(k+1)}) - (A\_compass_{(k)} - A\_Gyro_{(k)}) \quad (6)$$

where: $A\_compass_{(k+1)}$, $A\_Gyro_{(k+1)}$, $A\_compass_{(k)}$, $A\_Gyro_{(k)}$ are the orientations of compass and MEMS gyroscope at time k + 1 and time k respectively.

Δt—time interval from time k to time k + 1.

H_{k+1}—measurement matrix of the STKF, which can be calculated by the measured temperature and the rate of temperature variation.

$$H_{k+1} = \begin{bmatrix} T_{k+1} & T'_{k+1} & 1 \\ T_k & T'_k & 1 \\ T_{k-1} & T'_{k-1} & 1 \end{bmatrix}.$$

As the initial values of H_{k+1}, H_1, H_2 and H_3 is measured when the gyroscope is motionless in the beginning.

T_{k+1}—measured temperature of the gyroscope at time k + 1.

T'_{k+1}–rate of temperature variation at time k + 1; for ease of use, it is simplified as: $T'_{k+1} = (T_{k+1} - T_k) / \Delta t$.

The parameters of bias drift model change with the environment, so if the environment doesn't change too much, the parameters also don't change sharply. Then we can make an assumption that the parameters are unchanged in short time, so the matrices Φ(k+1,k) and Γ(k+1,k) are all identity matrix.

Because of the inaccuracy of the model and noise characteristics, the correction effect of the new measurements on the estimated value decreases with time, the correction effect of the early measurements increases, and the error of the estimated value of the traditional KF increases. Thus, the STKF is used to increase the weight of the new measurements to avoid the error of the estimated value increase with time.

Increasing the variance of the measurement noise *R* and the initial value of the covariance matrix P_0 can decrease the influence from early measurements on the latest state variable; thus, the noise characteristics are necessary to be changed.

The characteristics of the measurement noise are modified as:

$$\left.\begin{aligned} E[v_k^N] &= 0 \\ Cov[v_k^N (v_k^N)^T] = R_k^N &= R_k * s^{N-k} (s > 1) \end{aligned}\right\} \tag{7}$$

where:

R_k^N —modified variance of the measurement noise at time *k*, which is used to calculate the state variable x^N at time *N*.

R_k—original variance of measurement noise at time *k*.

$\hat{x}^N$ —the estimation of state variables at time *N*, which recursively obtained from x^1 to $\hat{x}^{N-1}$, so the measured values y^1 to y^N are all useful to calculate the state variable $\hat{x}^N$; therefore, $\hat{x}^N$ is a linear combination of the measured values, and RN1 to RNN are all useful in calculating $\hat{x}^N$. If the characteristics of the measurement noise in Equation (7) are used, the longer the time interval span from time *k* to time *N*, the larger the value of R_k^N is, and the relevant gain matrix Kk becomes smaller. This means that the earlier time implies that the weight of measurement used for calculating $\hat{x}^N$ is smaller; therefore, y^k contributes less to $\hat{x}^N$.

The initial state characteristic is modified as:

$$\left.\begin{aligned} E[x_0^N] = m_{x_0} \\ Var[x_0^N] = P_0^N = P_0 * s^N \, (s > 1) \end{aligned}\right\} \tag{8}$$

The processing noise characteristic is modified as:

$$\left.\begin{aligned} E[\omega_{k-1}^N] = 0 \\ Cov[\omega_{k-1}^N (\omega_{k-1}^N)^T] = Q_{k-1}^N = Q_{k-1} * s^{N-k} \, (s > 1) \end{aligned}\right\} \tag{9}$$

The noise characteristic of the traditional KF is replaced by Equations (7)–(9) to obtain the STKF:

$$\left.\begin{aligned} \hat{x}_{k/k-1}^* &= \Phi_{(k,k-1)} \hat{x}_{k-1}^* \\ \hat{x}_k^* &= \hat{x}_{k/k-1}^* + K_k^* (y_k - H_k \hat{x}_{k/k-1}^*) \\ K_k^* &= P_{k/k-1}^* H_k^T (H_k P_{k/k-1}^* H_k^T + R_k)^{-1} \\ P_{k/k-1}^* &= \Phi_{(k,k-1)} (s P_{k-1}^*) \Phi_{(k,k-1)}^T + Q_{k-1} \\ P_k^* &= (I - K_k^* H_k) P_{k/k-1}^* \end{aligned}\right\} \tag{10}$$

Equation (10) shows that the difference between the STKF and the traditional KF is only the attenuation factor s. When $s > 1$, $P_{k/k-1}$ is greater than that of the traditional KF, and Kk becomes larger as well. Thus, $\hat{x}_k = (1 - K_k H_k) \hat{x}_{k/k-1} + K_k y_k$ shows that the influence from the old measurements on the estimated value is decreased, and the weight of the new measurement increases [23,24].

The above approach estimates the slow-changing component of the MEMS gyroscope accurately, and the parameters are unaltered in a short period of time; therefore the STKF is implemented at a certain frequency. A higher frequency will yield better results for the STKF, however, the noise of the compass causes errors in $\Delta A_{(k+1,k)}$ and also causes the measured value of the STKF to contain the error. If the frequency decreases, the time interval Δt of the STKF will increase, and the impact of the compass noise will be reduced as a result of $y_{k+1} = \hat{z}_k + \Delta A_{(k+1,k)} / \Delta t$. Considering the two abovementioned aspects, the interval for STKF implementation is set every 10 s. Because the environment does not change significantly in 10 s, the parameters are not altered significantly in that time, which can guarantee accuracy. Also the long time interval reduces the amount of calculation.

4. EXPERIMENTAL RESULTS

This section presents the results of the two temperature experiments. For ease of implementation, the first experiment was operated in the laboratory with a large change of temperature and in the second experiment, the system was installed on a vehicle to test its performance when the vehicle moved in an outdoor environment.

4.1. The Static Experiment

In the first experiment, the system is motionless during the entire process, and the temperature is increased from 23 °C to 58 °C at a non-uniform speed and then lowered to test the performance of this algorithm. The magnetic interference can be identified via the insensitivity of the MEMS gyroscope to magnetic interference; thus, the STKF is only carried out under the condition in which there is no magnetic interference.

Figure 1 shows that when the gyroscope temperature increased from 23 °C to 58 °C and then lowered to ambient temperature, the motionless gyroscope bias drifted with temperature. The gyroscope signal changed rapidly when the temperature varied quickly. Thus, the error of the gyroscope accumulated rapidly if the bias was not compensated accurately.

The blue line in Figure 2 is the uncompensated angular velocity of the motionless gyroscope, the red line is the estimation of the gyroscope bias, and the green line is the ideal bias. We can see that the bias estimation follows the ideal bias very well.

In Figure 3, the blue line is the uncompensated angular velocity of the gyroscope, and the red line is the compensated the angular velocity. It can be seen that the compensated angular velocity of the gyroscope is not affected by the temperature.

Figure 4 shows comparison of the heading error by using different algorithms in the entire process. The blue line is the heading error compensated by the adaptive algorithm, and the error is kept in the range from −0.6° to 0.4°. The red line shows the heading error compensated by Kalman Filter (KF), and the error is kept in the range from −1.5° to 0.5°. The green line is the heading error compensated by MLR. The parameters are

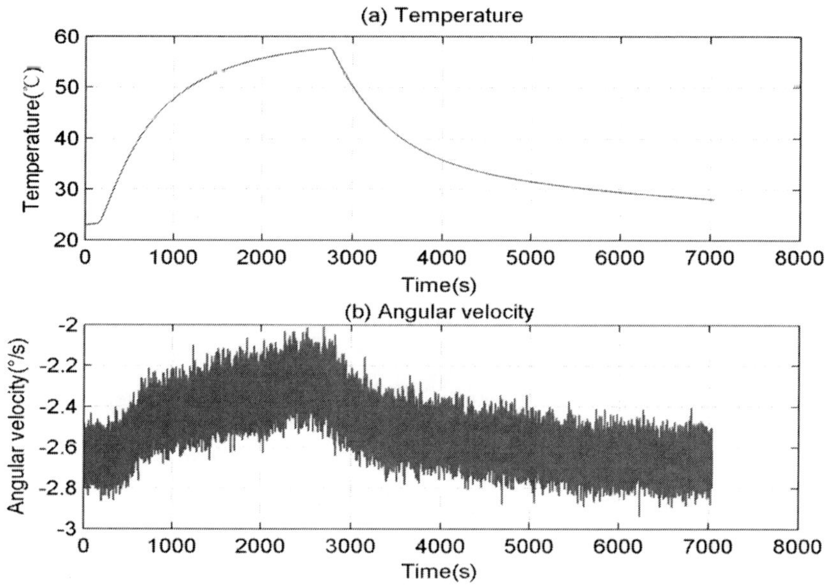

Figure 1. Original data of motionless gyroscope. (**a**) Temperature variation; (**b**) Angular velocity.

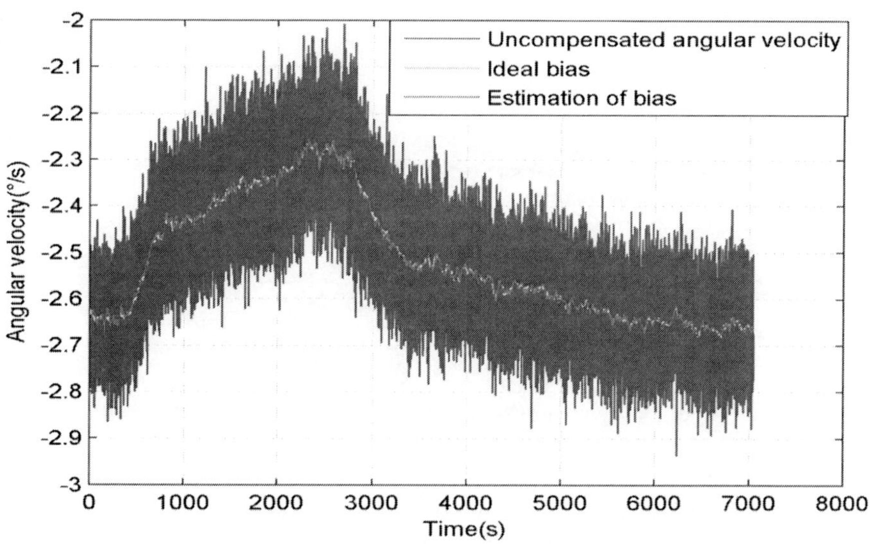

Figure 2. Uncompensated angular velocity and the estimation of the bias of the gyroscope.

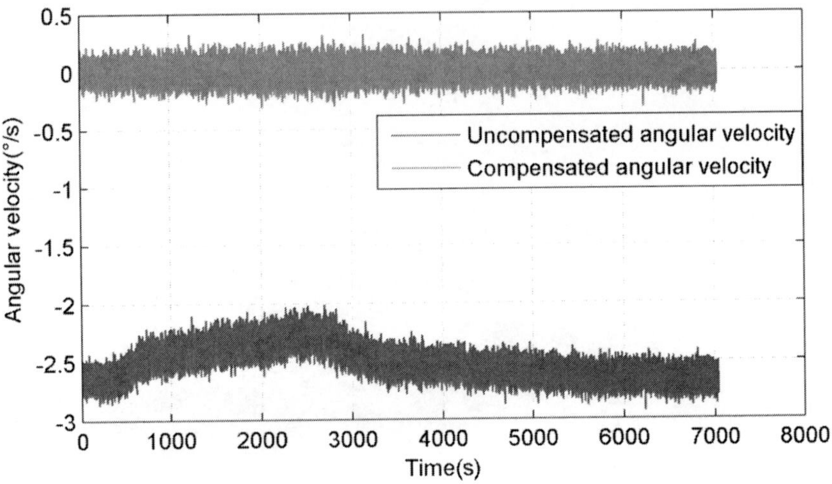

Figure 3. Uncompensated and compensated angular velocity of the gyroscope.

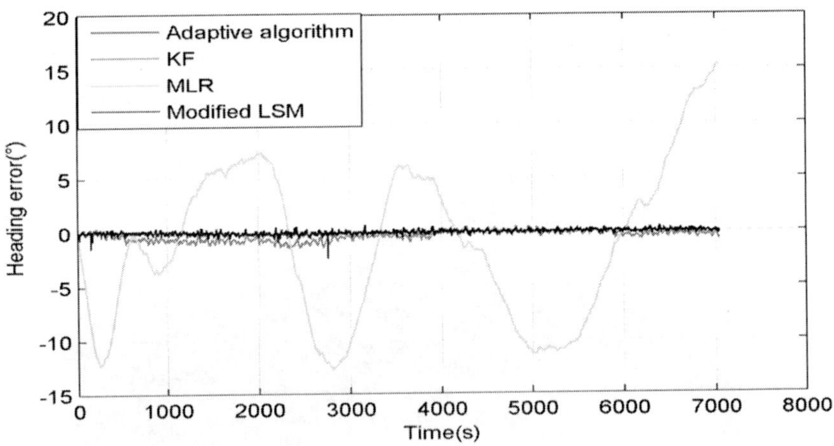

Figure 4. Heading error of different algorithms without interference.

calculated by the least squares method (called "LSM" for short) using data in the entire process. The heading error is much larger than the other methods, and the error is kept in the range from −12.5° to 15°. The black line is the heading error compensated by the modified LSM. It calculates the parameters every 10 s to make these three parameters change with environment and it almost overlaps with the blue line.

The adaptive algorithm does not work all the time, for example magnetic interference in the surroundings may cause some errors in the system, so this algorithm does not run to estimate the parameters. To test the temperature compensation performance of different algorithms when the algorithms do not work for some reason, we make the algorithms not work in three time periods (300 s, 300 s and 600 s). The results are shown in Figure 5. The blue line is the heading error which is compensated by the adaptive algorithm, where it can be seen that the algorithm also compensates the drift accurately when it is in the outage, and the maximum error is just less than 1.8°. The red line is the heading error which is compensated by KF, where the heading error increases significantly during the KF outage. The green line is the heading error which is compensated by MLR. The black line is the heading error which is compensated by the modified LSM. It shows that the adaptive compensation algorithm has a much higher precision than these three algorithms, when the algorithms are in outages for some reason.

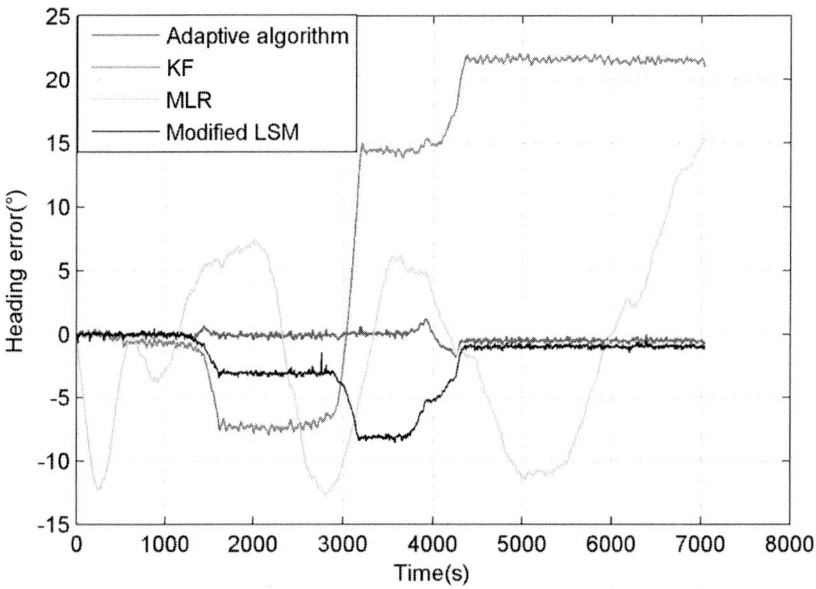

Figure 5. Heading error of different algorithms with interference.

On the basis of the experimental data, the Allan variance is used to analyse the performance of the algorithms, as shown in Figure 6. The blue line in Figure 6 indicates the Allan standard deviation of the uncompensated angular velocity error, and the red line indicates the Allan standard deviation of the angular velocity error which is compensated by the adaptive algorithm. The green line indicates the Allan standard deviation of the angular velocity error which is compensated by KF. The black line indicates the Allan standard deviation of the angular velocity error which is compensated by MLR. It can be seen that the rate random walk (RRW) and rate ramp (RR) are obviously decreased [25,26]. The comparison of these lines shows the adaptive algorithm and KF compensate the temperature drift significantly better than the MLR. If there is no magnetic interference, the KF and adaptive algorithm have a similar performance. This adaptive algorithm significantly decreases the influence from the temperature and makes the MEMS gyroscope much more accurate and stable. Figure 6 also shows that the noise, which was at approximately $\tau = 10$ s, is not significantly decreased, which is because the STKF is implemented every 10 s, this makes the parameters change every 10 s, and the errors caused by the noise of compass and gyroscope are also different every 10 s, therefore, this still needs further improvement.

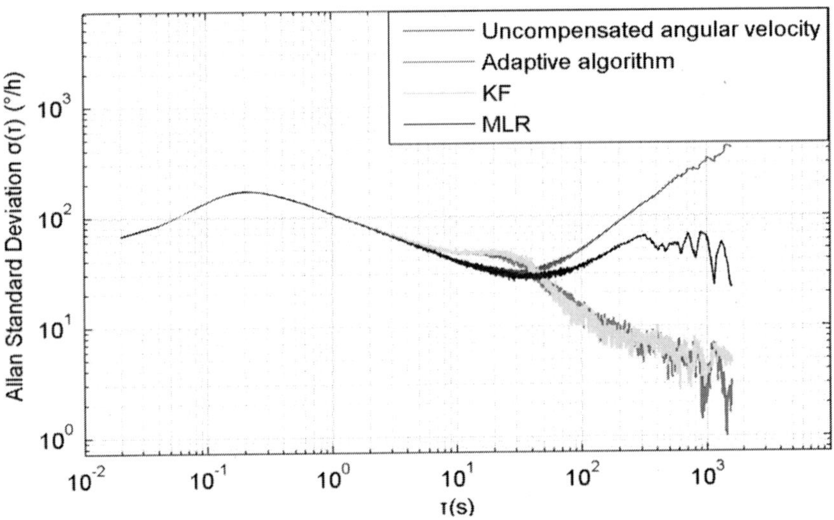

Figure 6. Allan standard deviation of uncompensated and compensated angular velocity error.

4.2. The Dynamic Experiment

After the static experiment, we installed the system on a vehicle, and moved outdoors to test the dynamic performance. To reduce the influence from pitching and rolling angles, the vehicle ran on a flat road. A fiber-optic gyroscope (FOG) (XW-GS1800) was fixed together with the system as the benchmark for the orientation. Figure 7 shows the temperature of the MEMS gyroscope in the dynamic experiment. The gyroscope temperature was influenced by the surrounding temperature, and changed from 26 °C to 37 °C.

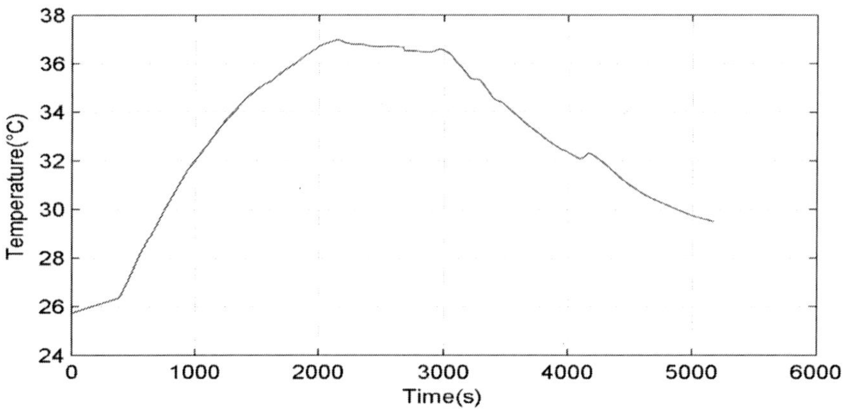

Figure 7. Temperature variation in the entire process.

The orientation in the entire process is shown in Figure 8. During most of the time, the vehicle moved around a rectangular area. The blue line is the compass orientation, the red line is the MEMS gyroscope orientation, and the green line is the FOG orientation. It can be seen that the MEMS gyroscope orientation followed FOG very well. The relative rotation angle of compass was not accurate, but when the system moved straight, and the heading error of the compass remained stable, so the adaptive algorithm only worked when the system moved straight.

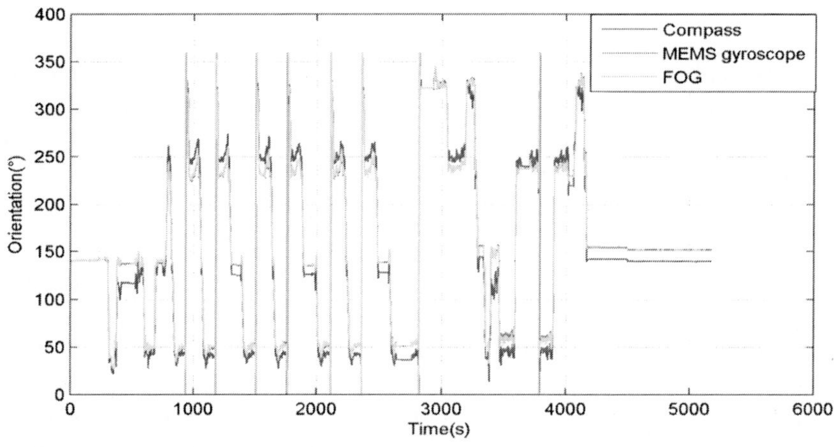

Figure 8. Orientation in the entire process.

The comparison of the results compensated by different algorithms is shown in Figure 9. The result of MLR is shown as a green line in Figure 9, and the heading error remains in the range from 15° to −13°. The KF improved the compensation precision to the range from 5° to −10°, and it is shown as a red line in Figure 9. The modified LSM didn't compensate the bias very well, because the measurement error had a great influence on its precision, as shown as a black line in Figure 9. The compensation precision of the adaptive algorithm is better than the former algorithms. Its heading error is shown as the blue line, and it is kept in the range from 5° to −2°.

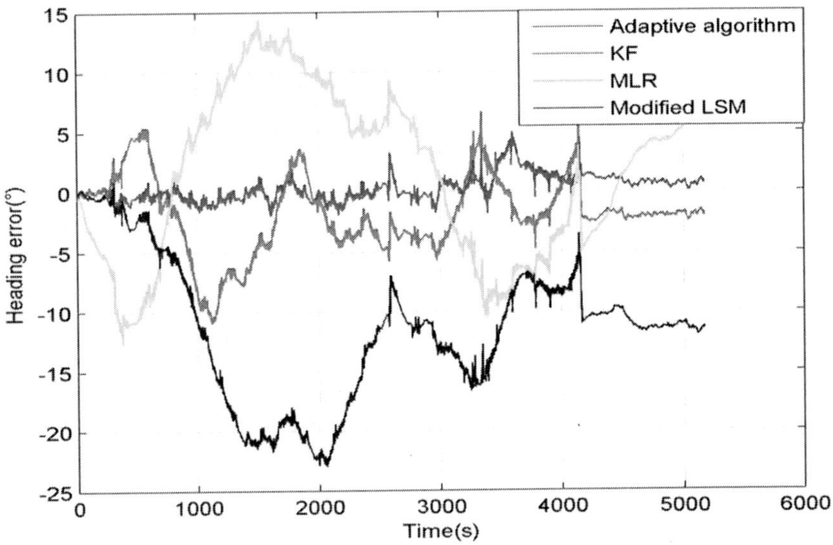

Figure 9. Heading error of different algorithms in the entire process.

These two experiments results demonstrate that this adaptive algorithm can compensate the temperature drift much more accurately than the KF, MLR and modified LSM. The adaptive algorithm estimates the parameters of the drift model. If the algorithm did not work for some reason such as magnetic interference, the parameters which were appropriate for the current environment could also maintain the high accuracy for a short time.

In Table 2, we compare the compensation results of the four methods in the experiment. In the static experiment, the compensation accuracy of MLR is not affected by interference. As shown in the table, the precision is improved to 0.002134°/s with the compensation by MLR. When there is no interference in the static experiment, the KF improved the compensation precision to -1.1496×10^{-4} °/s, while the modified LSM improved it to -2.4515×10^{-4} °/s, and the adaptive algorithm improved this precision to 2.1619×10^{-5} °/s. When interference existed in the static experiment, the adaptive algorithm also performed better than the other three methods, and the compensation precision is 1.7511 × 10^{-4} °/s. In the dynamic experiment, we used the FOG as the orientation benchmark to test the performance of these four methods. The mean of the heading error which was compensated by MLR is 1.4882°. The performance of KF and modified LSM were not as good as MLR, and they are −1.9085° and −11.5318°, respectively. The adaptive algorithm improved this precision to 0.2743°, and the heading error compensated by the adaptive algorithm was much more stable than with the other methods.

Table 2. Comparison of compensation results of the four methods.

		MLR	KF	Modified LSM	Adaptive Algorithm
Bias drift error in static experiment without interference	Mean (°/s)	0.002134	-1.1496×10^{-4}	-2.4515×10^{-4}	2.1619×10^{-5}
	Variance	5.5005×10^{-4}	7.5105×10^{-5}	1.1416×10^{-4}	6.7660×10^{-5}
Bias drift error in static experiment with interference	Mean (°/s)	0.002134	0.003023	-2.0156×10^{-4}	1.7511×10^{-4}
	Variance	5.5005×10^{-4}	3.4361×10^{-4}	1.8643×10^{-4}	9.3155×10^{-5}
Heading error in dynamic experiment	Mean (°)	1.4882	−1.9085	−11.5318	0.2743
	Variance	46.9166	9.8503	34.7090	0.9926

4.3. The Simulation of SF

The scale factor (called "SF" for short) is as important as bias drift in actual use, so it must be compensated accurately. There still are some unsolved problems for us to adaptively compensate both SF and bias drift in actual use, so we just use the data which is collected under the conditions with stable temperature and magnetic field to reduce the influence from magnetic interference. This data is combined with the data of bias and temperature in the static experiment. Then we generated the data of SF base on the information of this gyroscope datasheet. The compensation methods for bias and SF are all the adaptive algorithm. We used the static data to compensate the bias drift, and used the rotational data to compensate SF. The temperature in the simulation is same as in the static experiment.

The SF is shown in Figure 10. The blue line is the ideal SF, The green line is the estimation of SF which is calculated with fixed parameters. The red line is the adaptive estimation of SF. It can be seen that the adaptive estimation of SF follows the ideal SF very well.

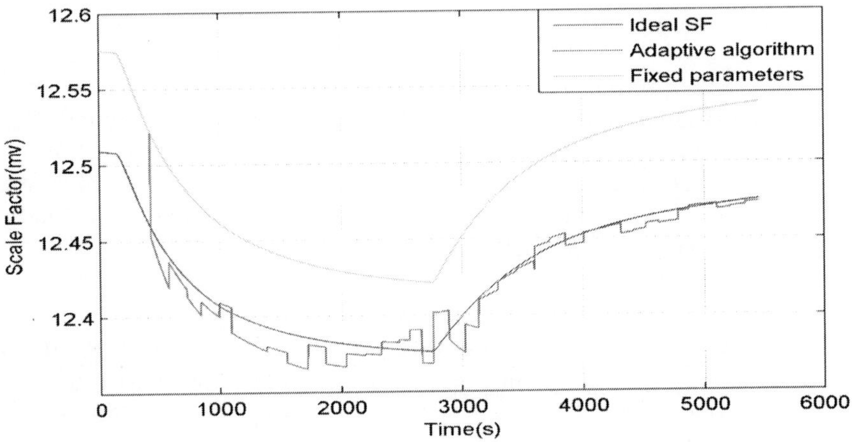

Figure 10. The SF in simulation.

The heading error in the entire process is shown in Figure 11. The blue line is the heading error without SF compensation. The green line is the heading error which is compensated by SF with fixed parameters. The red line is the heading error which is compensated by the adaptive estimation of SF. The reason of the heading error spikes during rotation is that the sample frequencies of the MEMS gyroscope and FOG are different. Thus, a heading error spike will occur if the system has a high rotational speed. The adaptive algorithm estimated the suitable parameters of SF model, so that it could compensate the SF much more accurately than the fixed parameters methods.

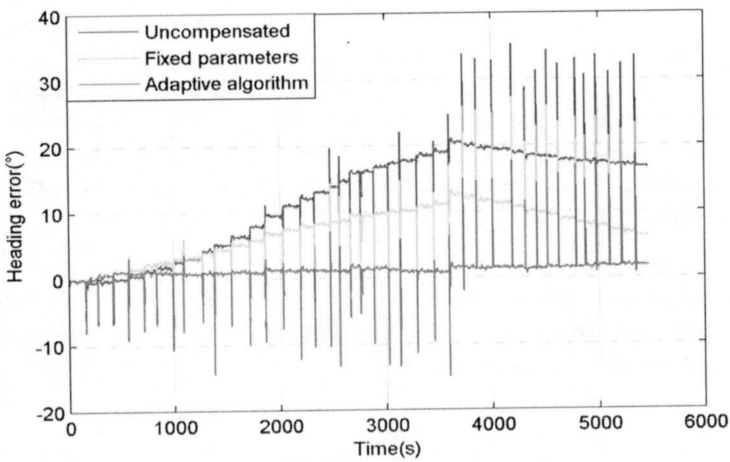

Figure 11. The heading error in simulation of SF.

Figure 10 and Figure 11 show that the SF has a great influence on the heading precision. The adaptive algorithm has a good performance in simulation, but before it is employed in actual use, there still are some problems that need to be solved, such as the fact the magnetic interference is difficult to identify during rotation, and the errors during rotation caused by SF errors or bias drift are difficult to identify too.

5. CONCLUSIONS

In this paper, we propose an adaptive algorithm for a system integrated with MEMS gyroscopes and a compass to compensate the bias drift of the MEMS gyroscope accurately in different environments and eliminate the influence of temperature variation. This algorithm only works in an environment without magnetic interference; otherwise, magnetic interference will introduce errors into the system. The proposed algorithm performs much better than KF and MLR, because it uses an STKF to estimate the parameters of the bias model of the MEMS gyroscope with the support of the compass orientation, and it changes these parameters to adapt to the environment intelligently, and then estimates the bias accurately using the model and the parameters. The significant characteristic of this algorithm is that when the algorithm is in an outage for some reason, these appropriate parameters can also maintain high precision to compensate the temperature drift.

This approach can be used easily, and the MEMS gyroscope does not need to calibrate its temperature model before use. As the state variables of the STKF, the parameters of the bias model change with the working environment. Thus, this algorithm has strong adaptability to different work environments and is easy to implement, so it is suitable for middle- or low-level inertial gyroscopes which always work under changing temperature conditions.

ACKNOWLEDGMENTS

This research was supported by National Natural Science Foundation of China (Project 61273082).

AUTHOR CONTRIBUTIONS

Yibo Feng defined the research theme, carried out most of the experiment and wrote the manuscript. Xisheng Li provided some guidance for the study and helped to write the manuscript. Xiaojuan Zhang helped to carry out the experiment and proofread the manuscript.

REFERENCES

1. Tanaka, M. An industrial and applied review of new MEMS devices features. *Microelectron. Eng.* **2007**, *84*, 1341–1344.
2. Fang, B.; Chou, W.S.; Ding, L. An Optimal Calibration Method for a MEMS Inertial Measurement Unit. *Int. J. Adv. Robot. Syst.* **2014**, *11*, 1–14.
3. Zhang, G.Y.; Deng, Z.L.; Fu, Z.X. Temperature modeling study for gyroscope. *J. Syst. Simul.***2003**, *15*, 369–378. (In Chinese).
4. El-Diasty, M.; El-Rabbany, A.; Pagiatakis, S. Temperature variation effects on stochastic characteristics for low-cost MEMS-based inertial sensor error. *Meas. Sci. Technol.* **2007**, *18*, 3321–3328.
5. Kirkko-Jaakkola, M.; Collin, J.; Takala, J. Bias prediction for MEMS gyroscopes. *IEEE Sens. J.***2012**, *12*, 2157–2163.
6. Zhao, L.; Wang, X.Z.; Cheng, J.H.; Huang, C. Research on the High Accuracy Temperature Control Box of FOG. In Proceedings of the 29th Chinese Control Conference, Beijing, China, 29–31 July 2010; pp. 3463–3467.
7. Zhao, X.; Su, Z.; Ma, X.F.; Tian, S.X. Research on bias compensation of MEMS gyroscope under large temperature difference. *Chin. J. Sens. Actuators* **2012**, *25*, 1079–1083. (In Chinese).
8. Zhao, X.H.; Yi, G.X.; Wang, C.H. Research on temperature compensation of silicon gyro's drift. *Transducer Microsyst. Technol.* **2008**, *27*, 48–50. (In Chinese).
9. Cheng, L.; Wang, S.R.; Ye, F. Research on bias temperature compensation for micromachined vibratory gyroscope. *Chin. J. Sens. Actuators* **2008**, *21*, 483–485. (In Chinese).
10. Zhou, C.L.; Zhang, Q.; Yan, S.H.; Gao, L.; Wang, G.C. Modeling and Compensation for Temperature errors of Interferometric Fiber Optic Gyroscope. In proceedings of the 2009 IEEE International conference on Information and Automation, Zhuhai/Macau, China, 22–25 June 2009; pp. 1443–1446.
11. Jin, J.; Wang, Z.; Zhang, Z.G.; Song, N.F.; Zhang, C.X. Temperature Errors Modeling for Fiber Optic Gyroscope Using Multiple Linear Regression Models. *J. Astronaut.* **2008**, *29*, 1912–1916.
12. Ding, J.C.; Zhang, J.; Huang, W.Q.; Chen, S. Laser Gyro Temperature Compensation Using Modified RBFNN. *Sensors* **2014**, *14*, 18711–18727.

13. Song, L.M.; Zhang, W. Temperature Error Compensation for Open-Loop Fiber Optical Gyro using Back-Propagation Neural Networks with Optimal Structure. In Proceedings of the 2014 IEEE Chinese Guidance, Navigation and Control Conference, Yantai, China, 8–10 August 2014; pp. 303–308.

14. Zhan, R.H.; Wan, J.W. Iterated Unscented Kalman Filter for Passive Target Tracking. *IEEE Trans. Aerosp. Electron. Syst.* **2007**, *43*, 1155–1163.

15. Zha, F.; Xu, J.N.; Li, J.S.; He, H.Y. IUKF neural network modeling for FOG temperature drift.*J. Syst. Eng. Electron.* **2013**, *24*, 838–844.

16. Shen, Z.; Georgy, J.; Korenberg, M.J.; Noureldin, A. Low cost two dimension navigation using an augmented Kalman filter/Fast Orthogonal Search module for the integration of reduced inertial sensor system and Global Positioning System. *Transp. Res. Part C* **2011**, *19*, 1111–1132.

17. Ren, H.L.; Kazanzides, P. Investigation of Attitude Tracking Using an Integrated Inertial and Magnetic Navigation System for Hand-Held Surgical Instruments. *IEEE/ASME Trans. Mechatron.* **2012**, *17*, 210–217.

18. Chen, X.Y.; Xu, Y.; Li, Q.H. Application of Adaptive Extended Kalman Smoothing on INS/WSN Integration System for Mobile Robot Indoors. *Math. Probl. Eng.* **2013**, *2013*, 1–8.

19. Li, W.; Wang, J.L. Effective adaptive Kalman filter for MEMS-IMU/magnetometers integrated attitude and heading reference systems. *J. Navig.* **2013**, *66*, 99–113.

20. Borenstein, J.; Ojeda, L. Heuristic drift elimination for personnel tracking systems. *J. Navig.***2010**, *63*, 591–606.

21. Fang, J.C.; Li, J.L.; Sheng, W. Improved temperature error model of silicon MEMS gyroscope with inside frame driving. *J. Beijing Univ. Aeronaut. Astronaut.* **2006**, *32*, 1277–1280. (In Chinese).

22. Won, S.-H.P.; Melek, W.W.; Golnaraghi, F. A Kalman/particle filter-based position and orientation estimation method using a position sensor/inertial measurement unit hybrid system. *IEEE Trans. Ind. Electron.* **2010**, *57*, 1787–1798.

23. Qin, Y.Y.; Zhang, H.Y.; Wang, S.H. *Theory of Kalman Filter and Integrated Navigation*, 1st ed.; NPU Press: Xi'an, China, 1998; pp. 88–92. (In Chinese)

24. Chen, X.Y.; Shen, C.; Zhang, W.B.; Tomizuka, M.; Xu, Y.; Chiu, K. Novel hybrid of strong tracking Kalman filter and wavelet neural network for GPS/INS during GPS outages. *Measurement* **2013**, *46*, 3847–3854.

25. Bekkeng, J.K. Calibration of a Novel MEMS Inertial Reference Unit. *IEEE Trans. Instrum. Meas.* **2009**, *58*, 1967–1974.

26. *IEEE Standard Specification Format Guide and Test Procedure for Single-Axis Interferometric Fiber Optic Gyros*; IEEE Standard 952–1997 (R2008); The Institute of Electrical and Electronics Engineers, Inc.: New York, NY, USA, 1997.

Chapter 9

A Robust Extended Kalman Filter for Speed-Sensorless Control of a Linearized and Decoupled PMSM Drive

P. Tety[1], A. Konaté[2], Olivier Asseu[2*], E. Soro[2], P. Yoboué[2], A. R. Kouadjo[2]

[1]*Institut National Polythechnique Houphouet Boigny (INPHB), Yamoussoukro, Côte d'Ivoire*

[2]*Ecole Supérieure Africaine des Technologies d'Information et de Communication (ESATIC), Abidjan, Cote d'Ivoire*

ABSTRACT

This paper uses a robust feedback linearization strategy in order to assure a good dynamic per- formance, stability and a decoupling of the currents for Permanent Magnet Synchronous Motor (PMSM) in a rotating reference frame (d, q). However this control requires the knowledge of certain variables (speed, torque, position) that are difficult to access or its sensors require the addi- tional mounting space, reduce the reliability in harsh environments and increase the cost of motor. And also a stator resistance variation can induce a performance degradation of the system. Thus a sixth-order Discrete-time Extended Kalman Filter approach is proposed for on-line estimation of speed, rotor position, load torque and stator resistance in a PMSM. The interesting simulations results obtained on a PMSM subjected to the load disturbance show very well the effectiveness and good performance of the proposed nonlinear feedback control and Extended Kalman Filter algorithm for the estimation in the presence of parameter variation and measurement noise.

Keywords: Robust Feedback Control, PMSM, Extended Kalman Filter, Estimation

1. INTRODUCTION

In recent years, there has been an emerging growth of PMSM. This machine has been widely used in many industrial applications. The main advantages, as compared with other AC motor drive, are high power factor, high power density, high torque to current ratio, high efficiency, Hence robustness, lower loss, lower maintenance and less complex motor can be obtained [1] [2] .

However, the control of PMSM is proved very difficult because the dynamic model of the PMSM is nonlinear, multidimensional and complex where some parameters vary with temperature or saturation. This nonlinear dynamic behavior induces the use of nonlinear feedback control strategy [3] [4] in order to permit a decoupling of the PMSM variables in a (d, q) coordinate so that stator currents can be separately controlled.

On the other hand, to preserve and improve the reliability under parameters variation and noises injected by the inverter (which can induce a state-space "coupling" and degradation of the system), a robust control approach has been made on the motor drives [5] - [7] . This control algorithm uses H-infinity synthesis of currents correctors in order to insure robust stability and performances of the inner current loop.

To guarantee good performances in presence of parameters variations (more specifically the stator resistance and load variation) and whereas advanced PMSM control strategies require knowledge of the instantaneous speed (which is difficult to access), the technique based on the state observer allowing an on-line estimation of the speed, position, load torque and the stator resistance is necessary.

Accurate estimation of speed in the presence of measurement and system noise, and parameter variations is a challenging task. Kalman filter (KF) named after Rudolph E. Kalman [8] is one of the most well known and often used tools for stochastic estimation. The KF is essentially a set of mathematical equations [9] [10] that implement a predictor-corrector type estimator that is optimal in the sense that it minimizes the estimated error covariance when some presumed conditions are met. For the speed, torque and stator resistance estimation problem of PMSM, where parameter variation and measurement noise is present, KF is the ideal one.

Many literatures on the KF technique and its applications, essentially extended for the estimation of the speed, have been published [11] -[13] . However, using the nonlinear feedback control, this Extended KF (EKF) technique doesn't take into account the combination load torque and stator resistance variation.

In the present research, after a brief review of the PMSM model, on the one hand a robust Input-output linearization and decoupling scheme is developed and on the other hand a sixth-order discrete-time EKF, based on KF principle, is proposed to estimate the speed, currents, position and extended for the load torque and stator resistance reconstruction.

Finally, the proposed combination nonlinear feedback control and EKF approach are confirmed by simulations results carried out on PMSM drive system in the presence of measurement noise and parameter variations.

2. PMSM EQUATIONS AND ROBUST FEEDBACK CONTROL

By assuming that the saturation of the magnetic parts and the hysteresis phenomenon are neglected; by considering the case of a smooth-air-gap PMSM (where the inductances are equal: $L_d = L_q$) and according to the field oriented principle where the direct axis current (I_d) is always forced to be zero which simplifies the dynamics and achieve maximum electromagnetic torque per ampere, the PMSM model in the rotor reference (d, q) frame are as follows [2] [5] :

$$\begin{cases} \dot{X} = F(X) + G \cdot U \\ Y = H(X) = \begin{bmatrix} h_1(X) & h_2(X) \end{bmatrix}^T = \begin{bmatrix} I_d & I_q \end{bmatrix}^T \end{cases} \tag{1}$$

with $X = \begin{bmatrix} I_d I_q \Omega \theta \end{bmatrix}^T$, $U = \begin{bmatrix} V_d V_q \end{bmatrix}^T$

$$G = \begin{bmatrix} \frac{1}{L_d} & 0 \\ 0 & \frac{1}{L_q} \\ 0 & 0 \\ 0 & 0 \end{bmatrix} = \begin{bmatrix} g_1 & 0 \\ 0 & g_2 \\ 0 & 0 \\ 0 & 0 \end{bmatrix}$$

This Equation (1) shows that the dynamic model of PMSM is nonlinear because of the coupling between the electrical currents and some parameters (in particular R_s) vary with temperature. Thus, in order to control independently the currents (I_d, I_q) and then preserve the robustness performance and stability of the system under parameters variation and measurement noise, we can uses a robust feedback linearization strategy to regulate the motor states [18] . Thus, we can see that the system (1) has relative degree $r_1 = r_2 = 1$ and can be transformed into a linear and controllable system by chosen:

- a suitable and an appropriate change of coordinates given by:

$$z_1 = h_1(x)\,;\; z_2 = h_2(x) \text{ with} \tag{2}$$

where $[v_1, v_2]^T$ are the new input vector of the obtained decoupled systems

- the feedback linearization control having the following form:

$$u = \begin{bmatrix} L_g h_1(x) & 0 \\ 0 & L_g h_2(x) \end{bmatrix}^{-1} \cdot \begin{bmatrix} v_1 - L_f h_1(x) \\ v_2 - L_f h_2(x) \end{bmatrix} = \begin{bmatrix} \frac{1}{L_q} & 0 \\ 0 & \frac{1}{L_d} \end{bmatrix}^{-1} \begin{bmatrix} v_1 - f_1 \\ v_2 - f_2 \end{bmatrix} \tag{3}$$

This feedback control (3) is effective and leads the system (1) to two decoupled subsystems;

- and two robust controllers C(s), using $H\infty$ synthesis and "Doyle method" [6] [14] , defined as:

$$C(s) = \frac{J(s)\cdot H(s)^{-1}}{1 - J(s)} \text{ with } J(s) = \frac{1}{(1 + t_0 s)^2} \text{ and } H(s) = \frac{1}{1 + Ts} \tag{4}$$

The real t_0 is an adjusting positive parameter, chosen adequately small ($t_0 < 1$), in order to satisfy the robustness performance, to have a good regulation and convergence of the currents.

The block diagram structure for the control of (I_d, I_q) can be summarized as follows in Figure 1.

However the control of a PMSM generally required the knowledge of the instantaneous speed of the rotor that is not measurable. Also a variation of the stator resistance or/and load torque ($R_s$and T_L) can induce a lack of field orientation. In order to achieve better dynamic performance, an on-line estimation of rotor speed, stator resistance and load torque is necessary.

In this study, in order to respect to the parametric variations in the presence of measurement and system noise (stochastic estimation), an EKF's algorithm for speed estimation extended for the stator resistance and load torque reconstruction, is presented and explained in the next section.

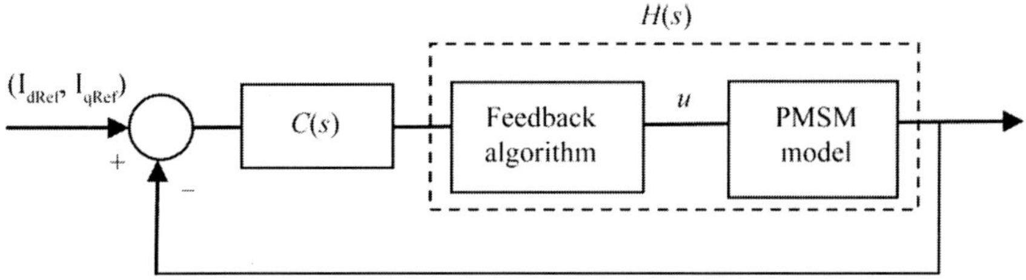

Figure 1. Proposed currents control scheme.

3. MODEL OF EXTENDED KALMAN FILTER

For parameter estimation using a full order EKF, the model structure (1) is discretized directly using Euler approximation (1st order) proposed in [15] . Furthermore, the state vector is extended to the stator resistance and load torque. Thus, choosing the currents (I_d, I_q), speed (Ω), rotor position (θ), load torque (T_L) and stator resistance (R_s) as state variables, the voltages (V_d, V_q) as inputs, the new discrete-time and Stochastic sixth-order nonlinear dynamic model for the PMSM is described by Equation (5):

$$\begin{cases} X_e(k+1) = f(X_e(k),U(k)) + n(k) = X_e(k) + T_e \cdot Q(x_e(k),U(k)) + n(k) \\ Y_e(k) = h(X_e(k)) + r(k) \end{cases} \tag{5}$$

where $X_e(k) = \left[I_d(k) I_q(k) \Omega(k) q(k) T_L(k) R_s(k)\right]^T$

, $U(k) = \left[V_d(k) V_q(k)\right]^T$

$$Q(X_e(k),U(k)) = \begin{bmatrix} -\frac{R_s(k)}{L_d} I_d(k) + \frac{L_q}{L_d} p \cdot I_q(k) \cdot \Omega(k) + \frac{1}{L_d} V_d \\ -\frac{R_s(k)}{L_q} I_q(k) - \frac{L_d}{L_q} p \cdot I_d(k) \cdot \Omega(k) - \frac{p \cdot \Phi_f}{L_q} \Omega(k) + \frac{1}{L_q} V_q \\ -\frac{f}{J} \cdot \Omega(k) + \frac{p \cdot \Phi_f}{J} I_q(k) - \frac{T_L(k)}{J} \\ p \cdot \Omega(k) \\ 0 \\ 0 \end{bmatrix}$$

$$Y_e(k) = \begin{pmatrix} I_d \\ I_q \end{pmatrix} = h(X_e(k)) = \begin{bmatrix} 1 & 0 & 0 & 0 & 0 & 0 \\ 0 & 1 & 0 & 0 & 0 & 0 \end{bmatrix} \cdot X_e(k) = H \cdot X_e(k)$$

The resulting output vector $Y_e(k)$ consists of the estimated motor current in a rotor reference frame being compared to the measured current. The difference is used to correct the state vector of the system model.

$X_e(k)$ and $Y_e(k)$ are the state vector and output, respectively at the k-th sampling instant, i.e. $t = k \cdot T_e$ with T_e the adequate sampling period chosen without failing the stability and the accuracy of the discrete-time model.

The random disturbance input, represented by $n(k)$, is the sum of modeling uncertainty, the discretization errors and the system noise. The measurement noise is represented by $r(k)$. Both $n(k)$ and $r(k)$, are assumed to be white Gaussian noise with zero mean and covariance matrix Q and R respectively. Consider that:

- $\hat{X}_e(k)$ = the estimate of $X_e(k)$ and $K(k+1) = \text{EKF}$ gain;
- $\hat{X}_e(k+1|k)$ = the linear minimum mean square estimate of $X_e(k+1)$;
- $P(k+1|k)$ = state prediction covariance error;
- $P(k+1|k+1)$ = state estimation covariance error;
- Initialization givens: $\hat{X}_e(0|0) = \hat{X}_e(0)$ and $P(0|0) = P(0)$.

The steps of the proposed sixth-order Discrete-time Extended Kalman Filter algorithm are as follows:

$$\begin{cases} 1)\ \hat{X}_e(k+1|k) = f\left(\hat{X}_e(k|k), U(k)\right) \\ 2)\ F(k) = \left.\dfrac{\partial f\left(\hat{X}(k), U(k)\right)}{\partial \hat{X}(k)}\right|_{X_e(k)=X_e(k|k)} \\ 3)\ P(k+1|k) = F(k)\cdot P(k|k)\cdot F^T(k) + Q \\ 4)\ K(k+1) = P(k+1|k)\cdot H^T\left[H\cdot P(k+1|k)\cdot H^T + R\right]^{-1} \\ 5)\ \Delta Y_e(k+1|k) = Y_e(k+1) - H\cdot \hat{X}_e(k+1|k) \\ 6)\ \hat{X}_e(k+1|k+1) = \hat{X}_e(k+1|k) + K(k+1)\cdot \Delta Y_e(k+1|k) \\ 7)\ P(k+1|k+1) = \left[I - K(k+1)\cdot H\right]\cdot P(k+1|k) \\ 8)\ \text{Increment } k \text{ and Go to step 1} \end{cases} \quad (6)$$

The EKF algorithm consists of repeated use of step (1-8) for each measurement. $F(k)$ is the Jacobian matrix of partial derivatives of $f(\bullet)$ with respect to $X_e(k)$. From Equation (5), we obtain:

$$F(k) = \begin{bmatrix} 1 & 0 & 0 & 0 & 0 & 0 \\ 0 & 1 & 0 & 0 & 0 & 0 \\ 0 & 0 & 1 & 0 & 0 & 0 \\ 0 & 0 & 0 & 1 & 0 & 0 \\ 0 & 0 & 0 & 0 & 1 & 0 \\ 0 & 0 & 0 & 0 & 0 & 1 \end{bmatrix} + T_e \cdot \begin{bmatrix} -\dfrac{R_s(k)}{L_d} + p\cdot\Omega(k) + p\cdot I_q(k) - \dfrac{I_d(k)}{L_d} + \dfrac{1}{L_d}V_d \\ -\dfrac{R_s(k)}{L_q} - p\cdot\Omega(k) - p\cdot I_d(k) - \dfrac{p\cdot\Phi_f}{L_q} - \dfrac{I_q(k)}{L_q} + \dfrac{1}{L_q}V_q \\ -\dfrac{f}{J} + \dfrac{p\cdot\Phi_f}{J} - \dfrac{1}{J} \\ p \\ 0 \\ 0 \end{bmatrix} \quad (7)$$

4. SIMULATION RESULTS AND DISCUSSION

The proposed robust nonlinear feedback control combined with a EKF strategy has been investigated with simulation tests carried out for a 1.6 kW PMSM by means of SIMULINK in order to illustrate its effectiveness against load, measurement noise and parameter variation (Figure 2). The nominal parameters of the PMSM, determined by means of the least-squares identification techniques proposed by the references [16] [17], are shown in the Table 1.

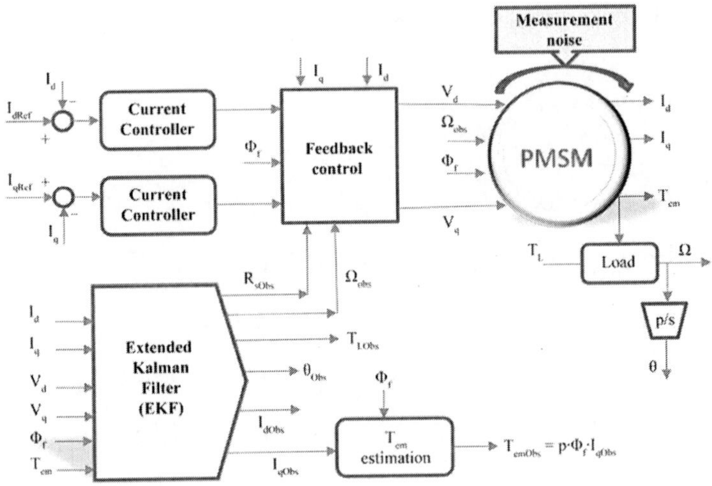

Figure 2. Simulation scheme.

Table 1. Nominal parameters of the PMSM.

$P_{mn} = 1.6$ kW	$U_n = 220/380$ V	$f_n = 0.0162$ N·m·sec·rad^{-1}
$p = 3$	$\Omega_n = 1000$ rpm	$J_n = 0.0049$ kg·m^2
$R_{sn} = 2.06\ \Omega$	$Phif_n = 0.29$ Wb	$L_{qn} = L_{dn} = 9.15$ mH

In order to evaluate its robustness and effectiveness, the comparisons between the estimated state variables and the simulated ones have been realized for several operating conditions with the presence of about 15% white noise on the measured currents and with additional load torque ($T_l > 1$ N·m).

Thus the simulations are obtained at first in the nominal case with the parameters of the PMSM (Table 1) and then in the second case, with 50% variation of the nominal stator resistance ($R_s = 1.5\ R_{sn}$) in order to verify the behavior of the proposed EFK algorithm estimator with respect to stator resistance and load torque variation.

Initialization and Tuning of the EKF Algorithms

The important and difficult part in the design of the full order EKF is choosing the proper values for the covariance matrices Q and R. The change of values of covariance matrices affects both the dynamic and steady-state.

In order to have a good performance, to insure better stability, convergence time and considerable rapidity of the EKF, the chosen values for the covariance matrices Q, R and P can be initialized and adjusted as follows:

$$P_{6\times6}(0) = diag\{10^4\};\ R_{2\times2} = diag\{10^3\};\ Q_{6\times6} = diag\{q\},$$

the real q_i must be tuned adequately small: $q \in [0 \quad 10^{-2}]$,

Our proposed Feedback control and EKF algorithm operate with a sampling period T_e = 1 ms and using Euler approximation.

Experiment simulation were performed and examined with regards to the following tasks: possibility current changing and load torque acting.

Figure 3 and Figure 4 show the responses of the currents, speed, rotor position, load torque and stator resistance for a step variation of the current reference (IdRef) under noisy conditions.

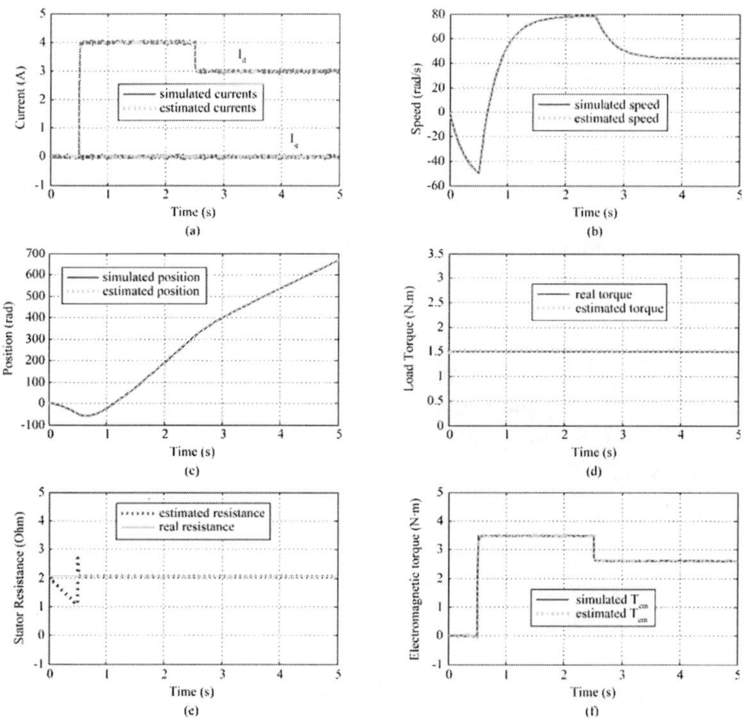

Figure 3. Nominal case ($R_r = R_{rn}$): Comparison between estimated and simulated values for T_L = 1.5 N·m in the presence of measurement noises.

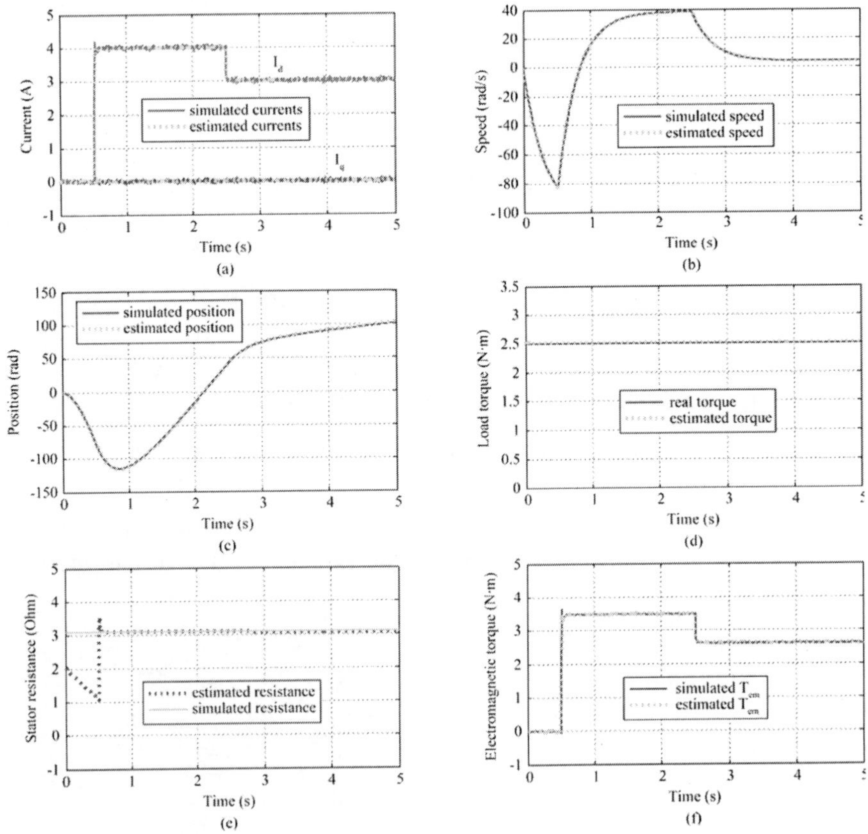

Figure 4. Non nominal case ($R_r = 1.5*R_{rn}$): Comparison between estimated and simulated values for T_L = 2.5 N·m in the presence of measurement noises.

One can see that in both nominal (Figure 3) and non-nominal cases (Figure 4 where R_r = 1.5 Rrn), the estimated values of currents, speed, rotor position and load torque converge very well to their simulated values and are not affected too much from the injected noise.

The observed better speed responses (Figure 3(b), Figure 4(b)), in the presence of parameter uncertainty and measurement noises, indicate the good regulation and convergence of the currents (with a decoupled system) due to a favorable stator resistance (Figure 3(e), Figure 4(e)) and load torque estimation (Figure 3(d), Figure 4(d)).

These good waveforms illustrate the fast convergence and high performance of the robust decoupling control and EKF algorithm against modeling uncertainty, parametric variation and measurement noise.

5. CONCLUSIONS

We have shown in this work that a robust nonlinear feedback control combined with a EKF approach have been realized to permit a linearization, decoupling and regulation of the PMSM states (currents) in order to assure a good dynamic performance of the global system and for solving range of problems in sensorless (speed, rotor position and load torque) control of PMSM drive without mechanical sensor.

The interesting simulation results obtained on the PMSM show the effectiveness, the convergence and the stability of the proposed control in presence of stator resistance variation, measured noise and load.

Thus in the industrial applications, because of the economic advantages (especially for low-powered motors), one will appreciate very well the experimental implement of this robust EKF sensorless control algorithm to substitute the PMSM mechanical sensor for the reconstitution of the speed, rotor position, load torque and stator resistance.

CITE THIS PAPER

P.Tety,A.Konate,OlivierAsseu,E.Soro,P.Yoboue,A. R.Kouadjo, (2015) A Robust Extended Kalman Filter for Speed-Sensorless Control of a Linearized and Decoupled PMSM Drive.*Engineering*,**07**,691-699.

REFERENCES

1. Dehkordi, A., Gole, A.M. and Maguire, T.L. (2005) PM Synchronous Machine Model for Real-Time Simulation. International Conference on Power Systems Transients, Montreal, Paper No. IPST05-159.
2. Pillay, P. and Krishnan, R. (1988) Modeling of Permanent Magnet Motor Drives. IEEE Transactions on Industrial Electronics, 35, 537-541.
3. Li, J. and Li, Y.Z. (2006) Speed Sensorless Nonlinear Control for PM Synchronous Motor Fed by Three-Level Inverter. IEEE International Conference on Industrial Technology, Mumbai, 15-17 December 2006, 446-451.
4. Marino, R., Tomei, P. and Verrelli, C.M. (2006) Nonlinear Adaptive Output Feedback Control of Synchronous Motors with Damping Windings. 32nd Annual Conference on IEEE Industrial Electronics, Paris, 6-10 November 2006, 1131- 1136.

5. Deng, Z.H. and Nian, X.H. (2015) Robust Control of Permanent Magnet Synchronous Motors. IEEE/CAA Journal of Automatica Sinica, 2, 143-150.

6. Bingyou, L. (2009) Research on H infinity Robust Tracking Controller for Permanent Magnet Synchronous Motor Servo System. International Conference on Information Engineering and Computer Science, ICIEC, 1-5

7. Zulkifli, S.A. and Ahmad, M.Z. (2011) H∞ Speed Control for Permanent Magnet Synchronous Motor. 2011 International Conference on Electronic Devices, Systems and Applications (ICEDSA), Kuala Lumpur, 25-27 April 2011, 290- 293.

8. Kalman, R.E. (1960) A New Approach to Linear Filtering and Prediction Problems. Journal of Basic Engineering, 82, 35-45.

9. Gowda, M., Ali, W., Cofie, P. and Fuller, J. (2013) Design and Digital Implementation of Controller for PMSM Using Extended Kalman Filter. Circuits and Systems, 4, 489-497.

10. Taibi, D., Titaouine, A., Benchabane, F. and Bennis, O. (2015) Stability Analysis of the Extended Kalman Filter for Permanent Magnet Synchronous Motor. Journal of Applied Engineering Science & Technology (JAEST), 1, 51-60

11. Aissa, A. and Mokhtari, B. (2012) Extended Kalman Filter for Speed Sensorless Direct Torque Control of a PMSM Drive Based Stator Resistance Estimator. Journal of Electrical and Control Engineering, 2, 33-39.

12. Fateh, B., Titaouine, A., Bennis, O., Guettaf, A., Yahia, K., et al. (2011) Robust Position and Speed Estimation Algorithm for Permanent Magnet Synchronous Drives. European Journal of Scientific Research, 57, 6-14.

13. Cheol, M., Hyun Nam, K., Jung, M., Chae, C. and Kwon, Y. (2012) Sensorless Speed Control of Permanent Magnet Synchronous Motor Using Unscented Kalman Filter. Proceedings of SICE Annual Conference (SICE), 2018-2023.

14. Doyle, J.C., Francis, B.A. and Tannenbaum, A.R. (1992) Feedback Control Theory. Maxwell MacMillan International, New York, 240 p.

15. Lewis, F. (1992) Applied Optimal Control Estimation-Digital Design and Implementation. Prentice Hall, New York, 448 p.

16. De Fornel, B. and Louis, J.P. (2007) Identification and Observation of the Electric Actuators. Hermes, Paris, 250 p.

17. Leite, V.T., Araujo, R.M. and Freitas, D.R. (2003) A New Online Identification Methodology for Flux and Parameters Estimation of Vector Controlled Induction Motors. IEEE International Electric Machines and Drives Conference, IEMDC'03, 1, 449-455.

18. Asseu, O., Yeo, Z., Koffi, M., Ori, T.R., Loum, G.L., Zoueu, T.J. and Tanoh, A. (2009) Reduced-Order Sliding Mode Flux Observer and Nonlinear Control of an Induction Motor. Trends in Applied Sciences Research, 4, 14-24.

Chapter 10

Application of the Multimodel Ensemble Kalman Filter Method in Groundwater System

Liang Xue

Department of Oil-gas Field Development Engineering, College of Petroleum Engineering, China University of Petroleum; No. 18 Fuxue Road, Changping District, Beijing 102249, China; Tel.: +86-10-8973-1629

ABSTRACT

With the development of *in-situ* monitoring techniques, the ensemble Kalman filter (EnKF) has become a popular data assimilation method due to its capability to jointly update model parameters and state variables in a sequential way, and to assess the uncertainty associated with estimation and prediction. To take the conceptual model uncertainty into account during the data assimilation process, a novel multimodel ensemble Kalman filter method has been proposed by incorporating the standard EnKF with Bayesian model averaging framework. In this paper, this method is applied to analyze the dataset obtained from the Hailiutu River Basin located in the northwest part of China. Multiple conceptual models are created by considering two important factors that control groundwater dynamics in semi-arid areas: the zonation pattern of the hydraulic conductivity field and the relationship between evapotranspiration and groundwater level. The results show that the posterior model weights of the postulated models can be dynamically adjusted according to the mismatch between the measurements and the ensemble predictions, and the multimodel ensemble estimation and the corresponding uncertainty can be quantified.

Keywords: Groundwater modeling; Data assimilation; Bayesian model averaging; The ensemble Kalman filter

1. INTRODUCTION

Groundwater resources are crucial to the development of human society and the sustainability of the environment. The rational management and effective protection of groundwater require accurate characterization of hydrogeological conditions. Due to the complexity of the groundwater system, various uncertainties exist and limit our ability to wisely manage the groundwater resources [1,2]. The uncertainty associated with the groundwater system mainly originates from the spatial heterogeneity of the hydrogeological properties. It can be quantitatively described by using geostatistical methods, with the capability to constrain the spatial distribution of hydrogeological properties on direct measurements [3]. To further improve the characterization of hydrogeological conditions, inverse modeling methods have been introduced to constrain the hydrogeological characterization on the available measurements of the state variables, such as hydraulic head, flow rate and solute concentration, and reduce the corresponding uncertainty associated with the characterization [4,5,6,7,8,9].

There has been a growing tendency to take conceptual hydrological model uncertainty into account in the past decade. The existence of conceptual uncertainty has been demonstrated in geoscience [10]. If the conceptual uncertainty is neglected, two types of errors may occur: first, the prediction can be statistically biased due to the adoption of an invalid conceptual model; and second, the uncertainty and risk associated with the prediction can be underestimated due to insufficient sampling of the conceptual model space [11]. The Generalized Likelihood Uncertainty Estimation (GLUE) pioneered the methods that can account for conceptual model uncertainty explicitly [12,13,14]. A more statistically coherent multimodel analysis method is the Bayesian Model Averaging method [15,16,17,18]. The GLUE and BMA (Bayesian Model Averaging) methods are combined together by [19,20]. The Maximum Likelihood version of BMA (MLBMA) is proposed to incorporate the groundwater model calibration methods based on maximum likelihood estimation into the multimodel analysis framework more conveniently [11,21,22,23,24].

With the development of *in-situ* monitoring technology in groundwater management, the ensemble Kalman filter method (EnKF) has gained significant attention due to its capability to update the model parameter and state simultaneously in a sequential manner [25,26,27]. Recently, we developed a multimodel data assimilation method by embedding the traditional EnKF into the BMA framework [28]. The novel multimodel EnKF method offers all of the advantages of the traditional EnKF method, and can also take conceptual model uncertainty into consideration during the data assimilation process. The methodology of the multimodel EnKF has been thoroughly described and the accuracy of the proposed method is quantitatively justified using the statistical criteria in [28]. This work focuses on demonstration of the multimodel EnKF analysis procedures when applying them to a more practical case study. In this case, the meteorological, topographic and geological data are practically obtained from the Hailiutu River area in the Ordos Basin, China. Limited by the dataset of the detailed hydrogeological characterization and dynamical hydraulic head observation, we adopt a strategy similar to [29], which utilizes a synthetic hydraulic head observation data computed from a generated full heterogeneous reference field conditional on the available direct hydraulic conductivity measurements; therefore, this study is essentially synthetic. However, the benefits of this strategy are that: (1) the application of the proposed method can be easily extended to other practical cases by imitating the example in this research; (2) it can reflect available field conditions, such as meteorological, topographic and geological conditions, to the maximum extent; and (3) the performance of the proposed method to estimate the hydraulic conductivity field can be evaluated by comparing it against the generated reference hydraulic conductivity field.

The rest of this paper is arranged as follows. In Section 2, the multimodel EnKF method is briefly presented, and the study area, the formulation of the multiple models, and the method implementation are demonstrated in detail. In Section 3, the analysis results are discussed. The main conclusions are summarized in Section 4.

2. METHODOLOGY

2.1. Theoretical Background

The general multimodel ensemble Kalman filter method is detailed in [28]. Here, the implementation of the multimodel EnKF method is briefly introduced in a specialized manner for the groundwater modeling problem, in which the dynamic hydraulic head observations are assimilated to estimate the hydraulic conductivity field.

Suppose that a set **M** of N_k conceptual models, M={1,⋯,Mk,⋯,MNk}, are postulated to simulate the groundwater system. For a given model M_k, the data assimilation process consists of two steps: the forecast step and the update step.

For the convenience of the method presentation, N_e realizations of state vector, **s**, are used here, which includes hydraulic conductivity, hydraulic head, and predicted values at the measurement locations:

$$\left\{\mathbf{s}_1,\mathbf{s}_2,\cdots,\mathbf{s}_{N_e}\right\}=\begin{Bmatrix}\mathbf{Y}_1 & \mathbf{Y}_2 & & \mathbf{Y}_{N_e}\\ \mathbf{h}_1 & \mathbf{h}_2 & \cdots & \mathbf{h}_{N_e}\\ \mathbf{m}_1 & \mathbf{m}_2 & & \mathbf{m}_{N_e}\end{Bmatrix} \tag{1}$$

where Y=lnK is the natural logarithm of hydraulic conductivity *K*; **h** is the hydraulic head; and**m** is the predicted values at the measurement locations.

Usually, the hydraulic conductivity and hydraulic head in the state vector are updated simultaneously in the EnKF method. However, the updated hydraulic head and hydraulic conductivity are not consistent with each other if the nonlinearity of the groundwater model is strong. To obviate this issue, only the hydraulic conductivity is updated during each assimilation step, and the corresponding hydraulic head is predicted by:

$$\mathbf{h}_{i,t}^{f}=M_k\left(\mathbf{Y}_{i,t-1}^{u}\right)+\xi_{i,t}=M_k\left(\mathbf{Y}_{i,t}^{f}\right)+\xi_{i,t} \tag{2}$$

where ξ_t is the model prediction error, with the assumption that $E(\xi_t)=0$ and $E(\xi_t\xi_t^T)=Q_t$; the subscript *i* is the index for the realizations used in the EnKF, *i* = 1, 2, ..., N_e; the subscript *t* is the index for the assimilation step; and the superscripts *f* and *u* represent *forecast*and *update*, respectively. This equation indicates that the hydraulic head for the *i*th realization in the ensemble at time *t* is forecasted by the updated log hydraulic conductivity for the *i*th realization in the ensemble at time *t*−1 through groundwater simulation model Mk(·), and the updated log hydraulic conductivity at time *t*−1 is treated as the log hydraulic conductivity at time *t*in the forecast step. Together with the hydraulic head at time *t* and the observed hydraulic head at time *t*, they form the state vector $\mathbf{s}_{i,t}^{f}$.

The observation values and the true state vector at time *t* can be related through:

$$\mathbf{h}_{i,t}^{obs}=\mathbf{H}_t\mathbf{s}_t^{true}+\boldsymbol{\varepsilon}_t+\zeta_{i,t} \tag{3}$$

where the superscript *obs* represents the observed value; and the superscript *true* represents the true (usually unknown) status; ε_t is the Gaussian-distributed measurement error with $E\left(\boldsymbol{\varepsilon}_t\boldsymbol{\varepsilon}_t^T\right)=\mathbf{C}_{\mathbf{h},t}$; $\zeta_{i,t}$ is the added perturbation error [30], which is usually assumed to have the same distribution as the measurement error, *i.e.*, $\zeta_{i,t}\sim N(\mathbf{0},\mathbf{C}_{\mathbf{h},t})$; T is the transpose operator; and **H** is a matrix operator which only contains 0 and 1 elements, and can be written as:

$$\mathbf{H} = [\mathbf{0}\ \mathbf{I}] \tag{4}$$

where **I** is the identity matrix.

The ensemble mean and covariance in the forecast step are:

$$\overline{\mathbf{s}}_t^f = \frac{1}{N_e}\sum_{i=1}^{N_e} \mathbf{s}_{i,t}^f \tag{5}$$

$$\mathbf{C}_{\mathbf{s},t}^f = \frac{1}{N_e - 1}\sum_{i=1}^{N_e}\left\{\left[\mathbf{s}_{i,t}^f - \overline{\mathbf{s}}_t^f\right]\left[\mathbf{s}_{i,t}^f - \overline{\mathbf{s}}_t^f\right]^T\right\} \tag{6}$$

The Kalman gain at time *t* can be written as:

$$\mathbf{K}_t = \mathbf{C}_{\mathbf{s},t}^f \mathbf{H}_t^T \left(\mathbf{H}_t \mathbf{C}_{\mathbf{s},t}^f \mathbf{H}_t^T + \mathbf{C}_{\mathbf{h},t}\right)^{-1} \tag{7}$$

The updated ensemble member of the state vector is:

$$\mathbf{s}_{i,t}^u = \mathbf{s}_{i,t}^f + \mathbf{K}_t\left[\mathbf{h}_{i,t}^{obs} - \mathbf{H}_t \mathbf{s}_{i,t}^f\right] \tag{8}$$

The ensemble mean and covariance of the updated state vector, conditional on all of the available observation data up to time *t* and the given model, are:

$$E\left(\mathbf{s}_t^u \mid \mathbf{h}_{1:t}^{obs}, M_k\right) = \frac{1}{N_e}\sum_{i=1}^{N_e} \mathbf{s}_{i,t}^u \tag{9}$$

$$Cov(\mathbf{s}_t^u \mid \mathbf{h}_{1:t}^{obs}, M) = \frac{1}{N_e - 1}\sum_{i=1}^{N_e}\left\{\left[\mathbf{s}_{i,t}^u - E\left(\mathbf{s}_t^u\right)\right]\left[\mathbf{s}_{i,t}^u - E\left(\mathbf{s}_t^u\right)\right]^T\right\} \tag{10}$$

where $\mathbf{h}_{1:t}^{obs} = [\mathbf{h}_1^{obs}, \mathbf{h}_2^{obs}, \cdots, \mathbf{h}_t^{obs}]^T$ is the observed hydraulic head data up to time *t*.

Bayesian model averaging framework provides a coherent way to deal with the situation in which multiple conceptual models are postulated. It shows that the posterior distribution of the predictions, conditional on all of the available data, is:

$$p\left(\Delta \mid \mathbf{h}_{1:t}^{obs}\right)=\sum_{k=1}^{K} p\left(\Delta \mid \mathbf{h}_{1:t}^{obs}, M_k\right) p\left(M_k \mid \mathbf{h}_{1:t}^{obs}\right) \tag{11}$$

This equation indicates that the posterior distribution of the predictions after averaging the model space is the weighted summation of the posterior distribution based on individual models. Now the key issue is to evaluate the posterior model weight.

The posterior model weight can be computed through Bayes' theorem as:

$$p\left(M_k \mid \mathbf{h}_{1:t}^{obs}\right)=\frac{p\left(\mathbf{h}_{1:t}^{obs} \mid M_k\right) p\left(M_k\right)}{\sum_{l=1}^{K} p\left(\mathbf{h}_{1:t}^{obs} \mid M_l\right) p\left(M_l\right)} \tag{12}$$

The marginal likelihood can be evaluated through the Monte Carlo integration method as:

$$\begin{aligned} p\left(\mathbf{h}_{1:t}^{obs} \mid M_k\right) &= \int p\left(\mathbf{h}_{1:t}^{obs} \mid \mathbf{Y}_{k,t}^{u}, M_k\right) p\left(\mathbf{Y}_{k,t}^{u} \mid M_k\right) d\mathbf{Y}_{k,t}^{u} \\ &\approx \frac{1}{N_e} \sum_{i=1}^{N_e} p\left(\mathbf{h}_{i,1:t}^{obs} \mid \mathbf{Y}_{i,k,t}^{u}, M_k\right) \end{aligned} \tag{13}$$

Several types of likelihood functions can be found in [19]. Here, we use the form:

$$p\left(\mathbf{h}_{i,1:t}^{obs} \mid \mathbf{Y}_{i,k,t}^{u}, M_k\right)=\left[\left(\mathbf{h}_{i,k,1:t}^{u}-\mathbf{h}_{i,1:t}^{obs}\right)^T \mathbf{C}_{\mathbf{h}_{1:t}^{obs}}^{-1}\left(\mathbf{h}_{i,k,1:t}^{u}-\mathbf{h}_{i,1:t}^{obs}\right)\right]^{-E} \tag{14}$$

where E is a user-defined parameter (we choose $E = 1$ in this work).

The posterior mean and covariance of the estimated hydraulic conductivity, after model averaging, are, respectively:

$$E\left(\mathbf{Y}_{k,t}^{u} \mid \mathbf{h}_{1:t}^{obs}\right)=E_M E\left(\mathbf{Y}_{k,t}^{u} \mid \mathbf{h}_{1:t}^{obs}, M\right)=\sum_{k=1}^{N_K} E\left(\mathbf{Y}_{k,t}^{u} \mid \mathbf{h}_{1:t}^{obs}, M_k\right) p\left(M_k \mid \mathbf{h}_{1:t}^{obs}\right) \tag{15}$$

$$
\begin{aligned}
&Cov(\mathbf{Y}_{k,t}^{u} \mid \mathbf{h}_{1:t}^{obs}) \\
&= E_{M_k} Cov(\mathbf{Y}_{k,t}^{u} \mid \mathbf{h}_{1:t}^{obs}, M_k) + Cov_{M_k} E(\mathbf{Y}_{k,t}^{u} \mid \mathbf{h}_{1:t}^{obs}, M_k) \\
&= \sum_{k=1}^{K} Cov(\mathbf{Y}_{k,t}^{u} \mid \mathbf{h}_{1:t}^{obs}, M_k) p\left(M_k \mid \mathbf{h}_{1:t}^{obs}\right) \\
&+ \sum_{k=1}^{K} \left[E(\mathbf{Y}_{k,t}^{u} \mid \mathbf{h}_{1:t}^{obs}, M_k) - E(\mathbf{Y}_{k,t}^{u} \mid \mathbf{h}_{1:t}^{obs}) \right] \\
&\cdot \left[E(\mathbf{Y}_{k,t}^{u} \mid \mathbf{h}_{1:t}^{obs}, M_k) - E(\mathbf{Y}_{k,t}^{u} \mid \mathbf{h}_{1:t}^{obs}) \right]^{T} p\left(M_k \mid \mathbf{h}_{1:t}^{obs}\right)
\end{aligned}
\tag{16}
$$

The posterior mean and covariance of the predicted hydraulic head, after model averaging, can be obtained simply by replacing $\mathbf{Y}_{k,t}^{u}$ with $\mathbf{h}_{k,t}^{u}$ in Equations (15) and (16).

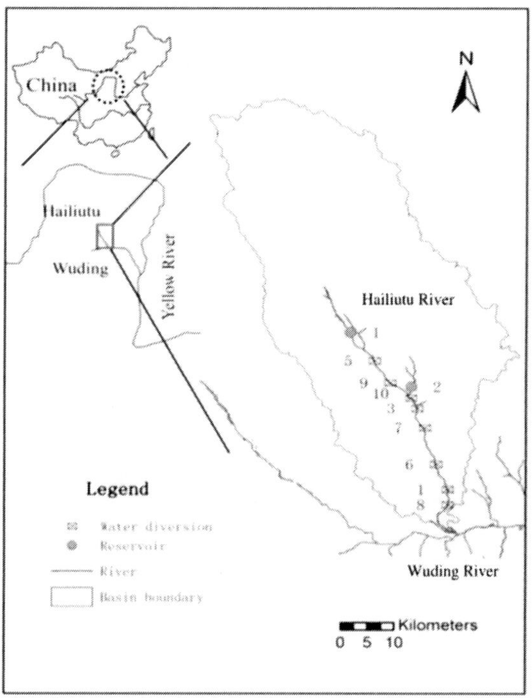

Figure 1. Location of the Hailiutu River Basin.

2.2. Study Area

The study site, the Hailiutu River Basin, is located in a semi-arid region of northwest China as shown in Figure 1, covering an area of about 2600 km² between 38°06' N and 38°50' N and 108°37' E and 109°15' E. Groundwater level fluctuation is mainly driven by seasonally varying recharge and evapotranspiration. Based on the geological investigation information, two major layers in the aquifer system are identified: the overlying unconfined layer consists of Quaternary Shala-Usun fine sand, with thickness between 0.5 and 217 m; and the underlying confined layer consists of Cretaceous Luohe sandstone, with thickness between 20 and 637 m (see the elevation of the studied aquifer system in Figure 2). No regional aquitard has been found between these two aquifer layers. The Hailiutu River flows mainly through the upper layer and cuts into the second layer near the end of the river system.

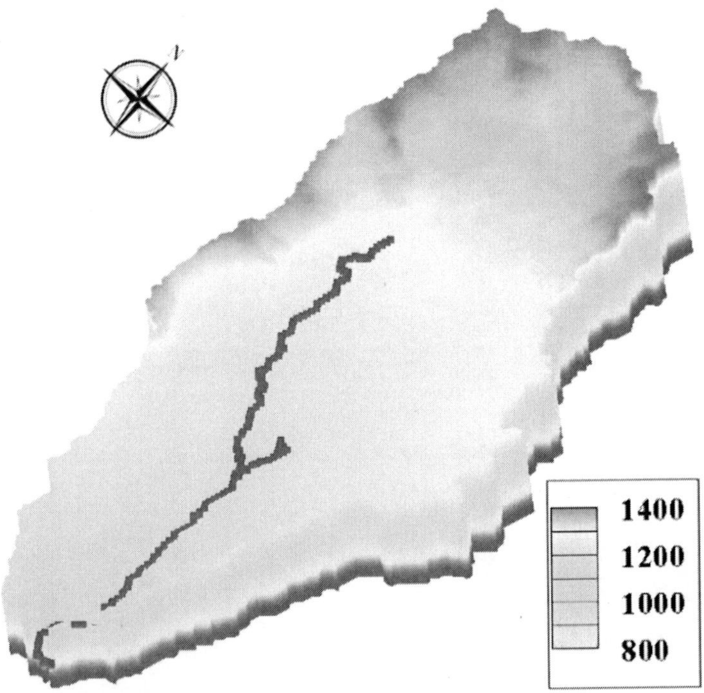

Figure 2. The elevation (in meters) contour of the studied aquifer.

2.3. Model Setup

The groundwater simulation software, MODFLOW, is used to build the numerical model of the study site. The study area is discretized into a 200 × 140 grid blocks with two layers, and each block represents a 500 m × 500 m area. The boundary of the study area is delineated based on the watershed divide, and it is treated as a no-flow boundary. The cells out of the boundary are characterized as inactive cells, as shown by the gray area in Figure 3. The number of the active cells is 20,840. The upper layer is an unconfined aquifer with a specific yield of 0.1, and the bottom layer is a confined aquifer with a specific storage of 1×10^{-5} m^{-1}. The discretized grid for each layer is shown in Figure 3. The yellow line in Figure 3 represents the location of the river. The riverbed conductance is taken to be 300 m^2/d for the part of the river flowing through the upper layer and 150 m^2/d for the part of the river flowing through the bottom layer. The initial condition of this groundwater model is taken as the simulation result at the end of year 2006, which is obtained from a previous base flow analysis model. The simulation period is 3 years (field measurements of precipitation and pan evaporation from the beginning of 2007 to the end of 2009 are used), which are divided into 36 stress periods with the time step as 1 day, *i.e.*, each stress period corresponds to a specific month. The differences of stress periods lie in the recharge and evapotranspiration. The precipitation (measured by rain gauge) and evaporation (measured by evaporation pan with a diameter of 20 cm) rate during the simulation period are depicted in Figure 4. The pan evaporation is converted to the potential or maximum evapotranspiration by multiplying a conversion constant. Based on a previous study, the conversion constant is set to be 0.3. The recharge coefficient of this area is set to be 0.32, and the evapotranspiration extinction depth is 4.6 m in this area.

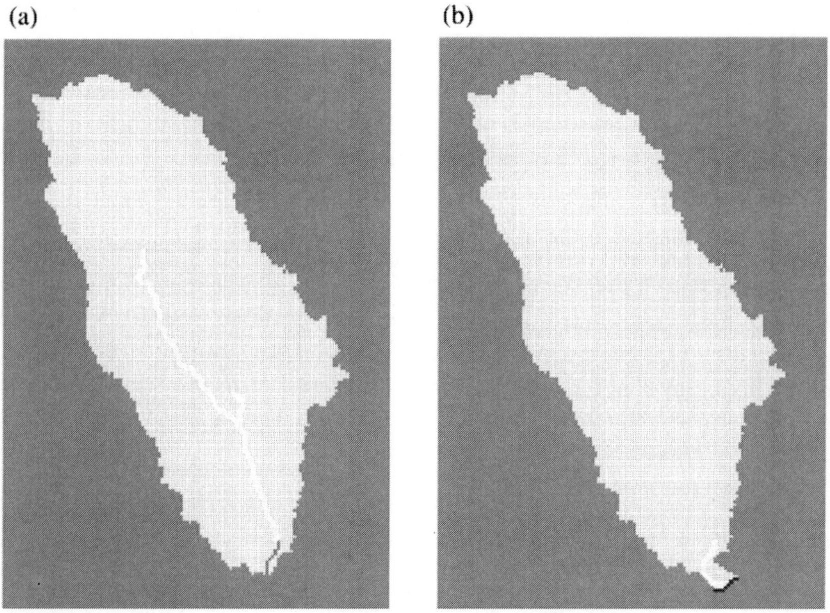

Figure 3. The discretized study area: (**a**) the upper layer; (**b**) the bottom layer. The yellow line represents the location of the river.

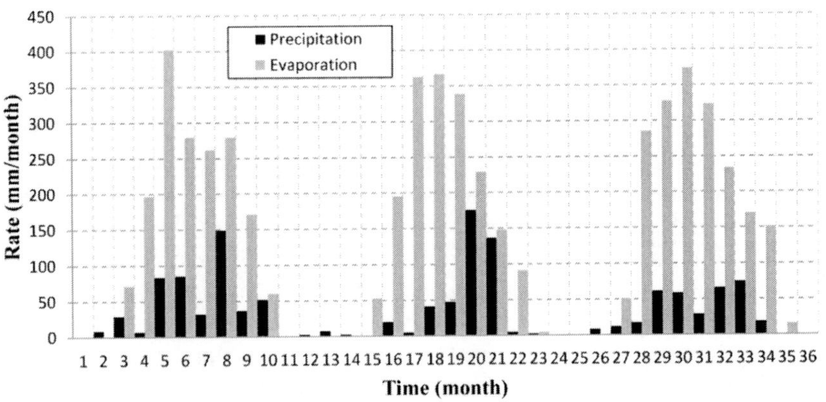

Figure 4. Precipitation (measured by rain gauge) and evaporation (measured by evaporation pan with a diameter of 20 cm) rates for 36 stress periods.

2.4. Synthetic Data Generation

Owing to the lack of sufficient and reliable real dynamic head observation data, a strategy similar to [29] is adopted, *i.e.*, a series of simulated head observation values are used for the analysis. The head observation values are computed based on the synthetically generated high-resolution log hydraulic conductivity reference field.

To capture the geological characterization of the field when generating the reference log hydraulic conductivity field by using the geostatistical method, the proper variogram model needs to be selected and the variogram parameters should be calibrated against the field measurements of the log hydraulic conductivity. The hydraulic conductivity measurements are listed in Table 1. It can be found that 10 hydraulic conductivity measurements are available for the upper layer. For a specific potential variogram model, the adjoint state maximum likelihood cross validation (ASMLCV) method proposed by [31] is employed to estimate the variogram model parameters of the upper layer of the aquifer. The ASMLCV minimizes the cross validation errors between the measured values and the Kriging estimated values, which is expressed in the likelihood function form:

$$NLL = M \ln 2\pi + \sum_{i=1}^{M} \ln \sigma_i^2 + \sum_{i=1}^{M} \frac{e_i^2}{\sigma_i^2} \tag{17}$$

where *M* is the number of the calibration data; *ei* is the cross validation error, which is the difference between the *i*th measurement and Kriged value using the rest of the measurements; and σ is the Kriging variance.

Three common variogram models are calibrated against the hydraulic conductivity measurements in the upper layer. They are exponential, Gaussian, and spherical variogram models, respectively. The proper model to describe the geostatistical property of the upper layer is selected based on the Kashyap information criterion (KIC) [22]:

$$KIC = -2NLL - 2\ln p\left(\hat{\boldsymbol{\theta}}\right) + N_k \ln\left(M / 2\pi\right) + \ln\left|\overline{\mathbf{F}}\right| \tag{18}$$

where $p\left(\hat{\boldsymbol{\theta}}\right)$ is the prior probability of the parameters $\hat{\boldsymbol{\theta}}$, in this case, the parameters are the sill and integral scale or range of the variogram models; N_k is the number of the parameters (here, it is 2); and $\left|\overline{\mathbf{F}}\right|$ is the determinant of the normalized Fisher information matrix. The lesser the KIC value is, the higher the model ranks. The estimated parameters, NLL, and KIC values are listed inTable 2. The exponential model is selected due to its lower KIC value. There are only 3 measurements for the bottom layer because the wells were not drilled deep enough; thus, the exponential variogram parameters for the bottom layer are inferred from a similar field in the Ordos Basin around the study area. Here, sill and integral scale of the bottom layer are set to be 0.15 and 20,000 m, respectively.

Table 1. The hydraulic conductivity measurements in the study area.

Well No.	x	y	Layer	*K* (m/d)
BH1	336,389.658	4,260,247.280	Upper	12.89
BH2	332,073.543	4,258,013.852	Upper	21.47
BH3	336,869.313	4,255,386.935	Upper	9.09
BH4	335,747.570	4,251,537.000	Bottom	0.39
BH5	333,485.000	4,251,550.000	Upper	17.92
BH7	331,812.363	4,243,332.846	Upper	14.65
BH8	337,568.173	4,246,870.834	Upper	17.12
BH9	323,502.996	4,239,096.890	Bottom	0.077
BH10	334,105.124	4,239,863.620	Upper	18.32
BH14	335,747.570	4,226,004.390	Upper	5.56
BH15	339,219.970	422,110.540	Bottom	0.066
BH16	324,070.370	4,240,827.500	Upper	10.15
BH17	342,183.141	4,252,717.435	Upper	15.64

The conditional (on all of the hydraulic conductivity data listed in Table 1) reference log conductivity field is generated using GSLIB [32]. To obtain the corresponding groundwater dynamic measurements, 25 synthetically designed head observation locations are used in this case. The generated log hydraulic conductivity field, head observation locations, and initial head distribution for each layer can be found in Figure 5.

Table 2. The parameter, NLL and KIC values, and model rank for the upper layer.

Model	Sill	Integral Scale/Range	NLL	KIC	Rank
Exponential	0.200	10,016.06	9.413	2.491	1
Gaussian	0.161	10,975.12	9.184	4.951	2
Spherical	0.211	27,543.90	9.281	6.071	3

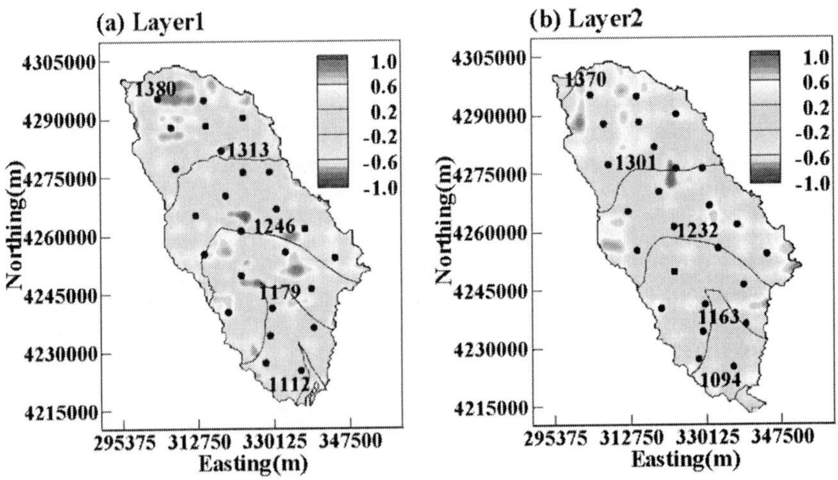

Figure 5. The synthetically generated high-resolution log hydraulic conductivity (flooded contour), initial hydraulic head (contour lines), and 25 synthetically designed head observation locations (black dots). (**a**) Layer 1; (**b**) Layer 2.

2.5. Multiple Conceptual Models

To construct multiple conceptual models to simulate this hydrogeological system, two uncertainty sources are considered. Geological model uncertainty due to different interpretations of geological and geophysical data is a well-known model uncertainty, and various zonations can be achieved by varying the number and distribution of hydraulic conductivity zones [33,34]. In particular, the hydraulic conductivity field in the two-layer aquifer system is partitioned into 6 and 8 zones as alternative geological models to the reference one with high resolution. The effect of evapotranspiration is crucial to groundwater management in semi-arid areas. The relationship between evapotranspiration rate and groundwater level is, however, uncertain. Since MODFLOW-2000, the evapotranspiration segments package (ETS) allows simulation of evapotranspiration with a user-defined relation [35]. A two-segment relation model (with the parameter PXDP = 0.4 and PETM = 0.3 at the intermediate segment endpoint) as the reference one is utilized to generate the head observations. As alternative models, a single-segment (linear, no intermediate segment endpoint) relation model and a three-segment model (with the parameters PXDP = 0.2 and PETM = 0.4 at the first intermediate segment endpoint and PXDP = 0.5 and PETM = 0.1 at the second one) are postulated in this case. The relationships used to describe the evapotranspiration rate and hydraulic head are depicted in Figure 6. The combination of two different zonations (with 6 and 8 zones, respectively) and two different segmented evapotranspiration relations (with 1 and 3 segments, respectively) gives 4 alternatively postulated simulation models. Note that the true two-segment evapotranspiration relation model is intentionally excluded from the postulated model set to reflect that the reference model is essentially unknown in practice.

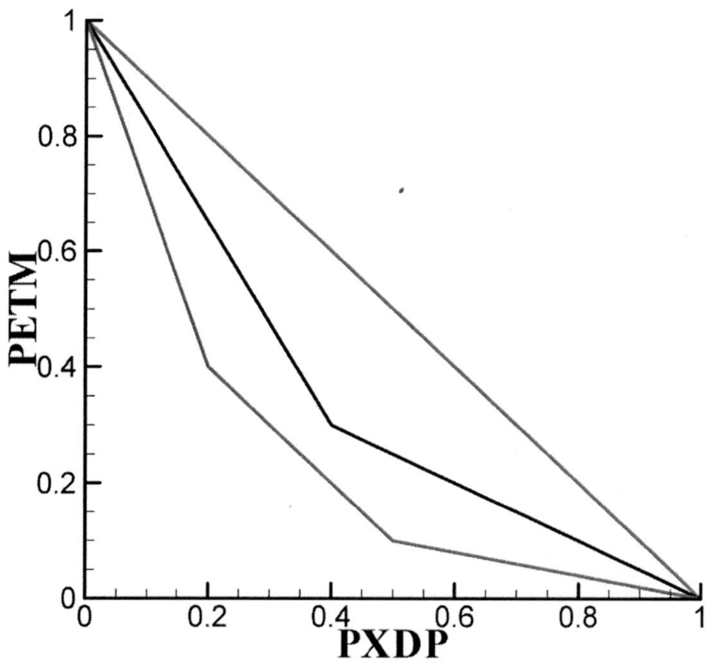

Figure 6. The reference relationship (black line) and two postulated relationships (red and blue lines) used in the ETS package.

The multimodel EnKF analysis procedures are the same as those presented in Section 2. The key parameters to implement the analysis are listed in Table 3.

Table 3. Summary of the implementation parameters.

Parameter	Value	Unit
Discretization		
Row	200	-
Column	140	-
Layer	2	-
Active cell for each layer	10,424	-
Grid spacing	500	m
Number of stress periods	36	-
Number of time steps	28–31	d
Flow (based on real field data)		
Horizontal hydraulic conductivity anisotropy ratio	1	-
Vertical hydraulic conductivity anisotropy ratio	0.1 (layer 1)	-
	0.01 (layer 2)	-
Specific yield for layer 1	0.1	-
Specific storage for layer 2	1×10^{-5}	m^{-1}
Riverbed conductance	300 (layer 1)	m^2/d
	150 (layer 2)	m^2/d
Maximum ET rate	0–0.004	m/d
ET extinction depth	4.6	m
Recharge flux	0–0.0018	m/d
Synthetic true hydraulic conductivity field		
Varigram model	Exponential	-
Mean	0	-
Variance	0.2 (layer 1)	-
	0.15 (layer 2)	-
Integral scale	10,000 (layer 1)	m
	20,000 (layer 2)	m
Measurement		
Number of head measurements	50	-
Standard deviation	10% of drawdown	m
The proposed multi-model EnKF		
Number of ensemble members	100	-
Number of assimilation steps	36	-
Number of postulated models	4	-
Label of the postulated models ("Z" and "S" represent zone and segment, respectively)	Z6S1	-
	Z6S3	-
	Z8S1	-
	Z8S3	-

3. RESULTS AND DISCUSSION

In this study, the assimilation step coincides with the stress period, *i.e.*, the hydraulic head measurements are assimilated to update the log hydraulic conductivity field at the end of each stress period. Figure 7 shows the dynamic change of posterior model weights in each assimilation step. In the first few steps, owing to the fact that less head observation information is available to be assimilated and the postulated geological models are both highly simplified through zonation (relative to the generated reference heterogeneous hydraulic conductivity field), the posterior weights of all the models are nearly uniformly assigned, e.g., each weight is nearly 25%. As more head observations become available, the posterior model weight of model Z8S3 tends to continue increasing and that of model Z6S1 exhibits the opposite tendency. For model Z6S3, the posterior model weight increases in the first 15 steps and starts decreasing after that. The change of the posterior model weight of model Z8S1 is oscillated without a general trend. After about 30 assimilation steps, these weights stabilize and the most sophisticatedly postulated model in this case, Z8S3, is ranked as the best model to describe the reference system, with a posterior model weight of 79.87%.

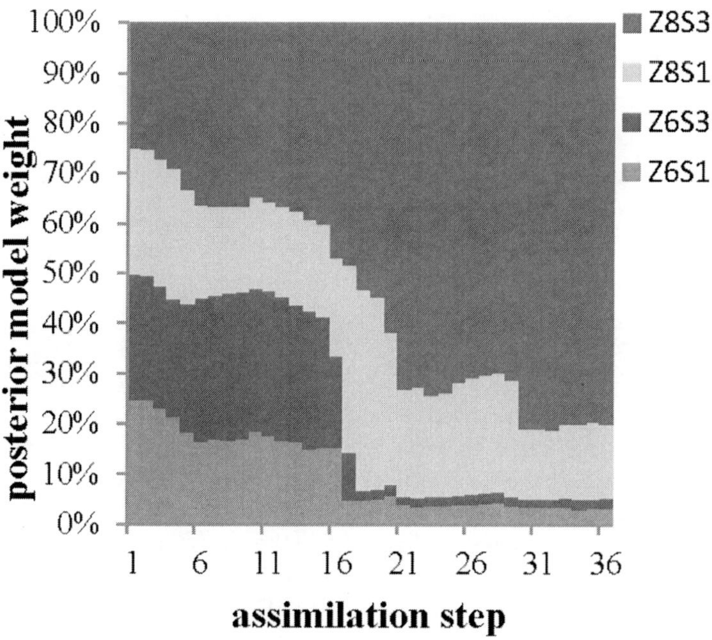

Figure 7. Posterior model weights changing with time for individual postulated models.

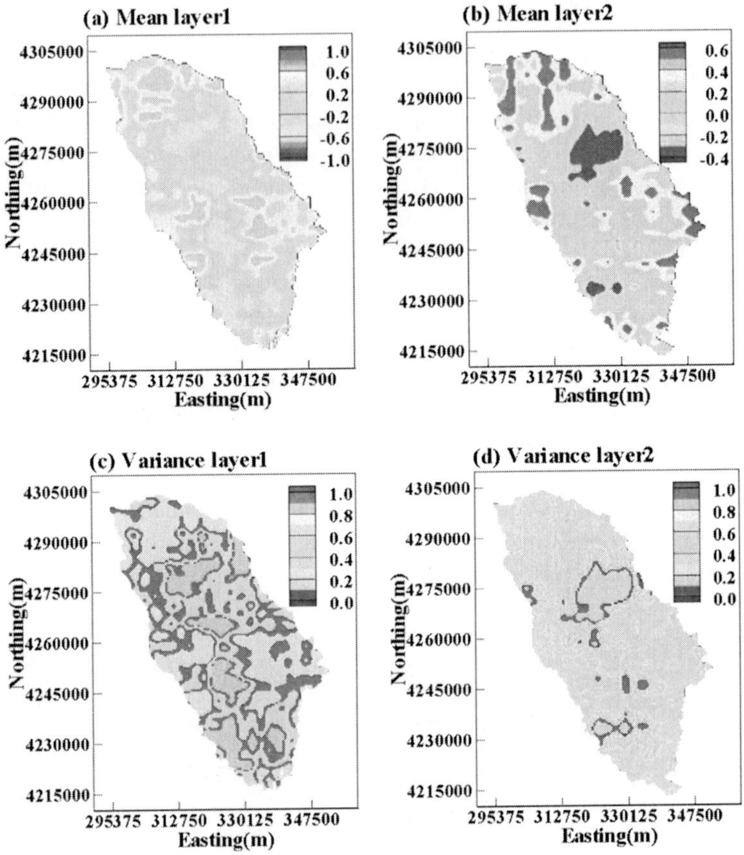

Figure 8. The multi-model ensemble mean and variance of log hydraulic conductivity for each layer before data assimilation.

Figure 8a,b and Figure 9a,b depict the multi-model ensemble mean for each layer before and after 36 steps of data assimilation in the studied aquifer system. It can be found that the ensemble mean after 36 steps of data assimilation captures the general pattern of the log hydraulic conductivity distribution by comparing them with the generated reference log hydraulic conductivity field shown in Figure 5. The corresponding multimodel ensemble estimation uncertainties before and after 36 steps of data assimilation for both layers are depicted in Figure 8c,d and Figure 9c,d. As expected, after the data assimilation process, the estimation uncertainty has been dramatically decreased. This quantified multimodel uncertainty is fundamentally more accurate than its counterpart based on the individual model because the conceptual model uncertainty can be taken into consideration, which can be an important uncertainty source. This type of uncertainty can propagate into the prediction of groundwater dynamics through the simulation of the groundwater system, and significantly affect the consequent decision-making or risk assessment based on the prediction. The conceptual model uncertainty has been ignored in most of the previous studies because no suitable algorithm can account for it. Here, with the proposed multimodel EnKF method, the overall uncertainty considering the conceptual model uncertainty can be thoroughly evaluated in each data assimilation step.

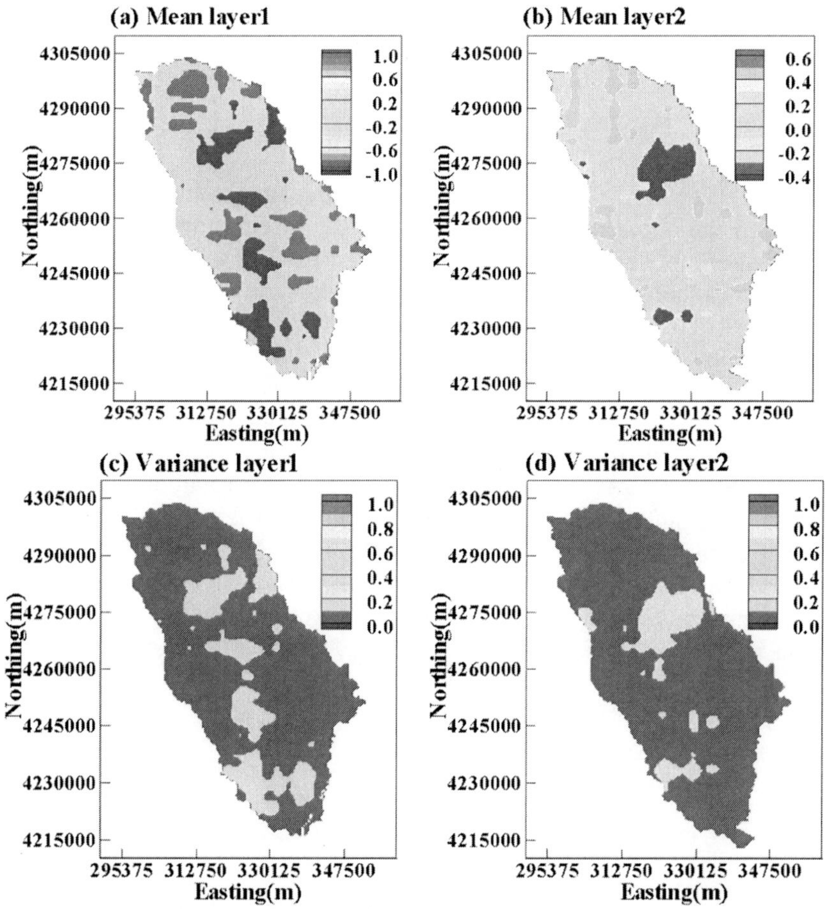

Figure 9. The multi-model ensemble mean and variance of log hydraulic conductivity for each layer after 36 steps of data assimilation.

To further investigate the uncertainty change in each assimilation step, the spatial averaged uncertainty or variance (SDV) value is analyzed. It is computed as:

$$\mathrm{SDV}(t) = \sqrt{\frac{1}{N_j}\sum_{j=1}^{N_j} Var(Y^u_{j,k,t} \mid \mathbf{h}^{obs}_{1:t})} \tag{19}$$

where N_j is the number of grid nodes; and j is the node index. It can be observed from Equation (16) that the total variance, $Var(\mathbf{Y}^u_{k,t} \mid \mathbf{h}^{obs}_{1:t})$, is composed of two parts: the within-model variance, $E_{M_k} Var(\mathbf{Y}^u_{k,t} \mid \mathbf{h}^{obs}_{1:t}, M_k)$ and the between-model variance, $Var_{M_k} E(\mathbf{Y}^u_{k,t} \mid \mathbf{h}^{obs}_{1:t}, M_k)$. The dynamic change of the spatial averaged variances, including total variance, within-model variance, and between-model variance, are plotted in Figure 10. If a full heterogeneous model is calibrated against the monitored hydraulic head data, it requires estimating 20,840 log hydraulic conductivity values, *i.e.*, log hydraulic conductivity values in all of the active cells, which may bring significant uncertainties associated with the estimation. However, with the zonation patterns in this case, it only requires estimating 6 log hydraulic conductivity values for model Z6S1 and Z6S3, and 8 log hydraulic conductivity values for model Z8S1 and Z8S3. Owing to the fact that the zonation dramatically reduces the number of parameters that need to be estimated, the within-model uncertainty decreases very rapidly with the assimilation of head observation data, and the between-model uncertainty increases in the beginning steps and then decreases when a certain amount of head data has been assimilated. The total uncertainty is dominated by the within-model uncertainty in the beginning and by the between-model uncertainty in the end. This highlights the importance of taking model uncertainty into account; otherwise, the risk induced by the uncertainty will be underestimated.

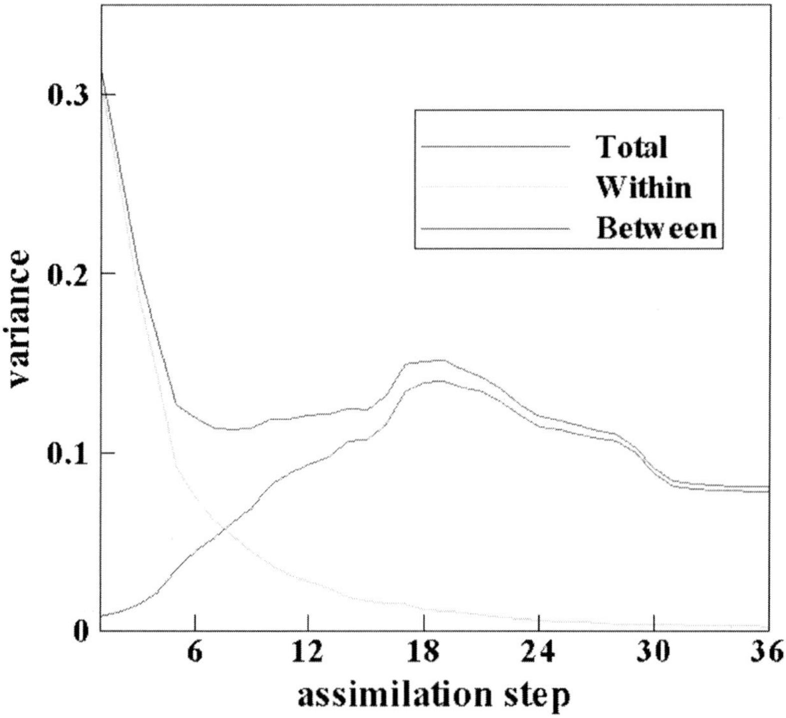

Figure 10. Dynamic changing of spatial averages of total multi-model ensemble variance, within-model variance, and between-model variance with time.

4. CONCLUSIONS

The proposed multimodel ensemble Kalman filter method is applied to a three-dimensional case study in this work to estimate the hydraulic conductivity field by assimilating dynamic hydraulic head observation data. The main contribution of the proposed multimodel EnKF method is that it can take conceptual model uncertainty into account and also possesses all of the capabilities of the widely used single-model based EnKF method. The accuracy and superiority of this method have been investigated in detail in a previous work published by the same author using a two-dimensional synthetic example. The aim of this paper is to show how to implement this method step-by-step in a more practical case. Despite the fact that part of the dataset is synthetically generated, it does not affect the demonstration of the application of this method in a practical case study. On the contrary, it helps us to assess the performance of this method explicitly while considering all of the collected field information.

Multiple conceptual models are constructed by considering two common but crucial uncertainty sources that influence the groundwater dynamics in a semi-arid study area, such as the Hailiutu River Basin, in this study. One is the zonation pattern of the hydraulic conductivity field, and the other is the relationship between the evapotranspiration and the groundwater level. The posterior model weight can be adjusted dynamically in each assimilation step, according to the mismatch between the predicted values and the actual observations. The mismatch is quantified by the likelihood function, and the posterior model weights are obtained through Bayes' theorem. The obtained multimodel ensemble mean is a weighted average of the ensemble means based on individual models, and the multimodel ensemble uncertainty consists of within-model uncertainty and between-model uncertainty. All of these terms can be quantified in each assimilation step. The results show that the posterior model weights change dynamically and tend to be stabilized when sufficient observations are assimilated. With the increasing of the assimilated information, the within-model uncertainty continues decreasing but the between-model uncertainty tends to increase in the beginning and stabilize in the late stage, and the overall uncertainty tends to decrease in the beginning and stabilize in the late stage.

ACKNOWLEDGMENTS

This work is partially funded by the National Natural Science Foundation of China (Grant No. 41402199), the Science Foundation of China University of Petroleum, Beijing (Grant No. 2462014YJRC038), and the Platform Construction Project for Researches on the Relationship between Water and Ecology in the Ordos Plateau (Grant No. 201311076).

AUTHOR CONTRIBUTIONS

As the sole author of this manuscript, Liang Xue developed the method, applied it to the case study, analyzed the simulation results, and wrote the paper.

REFERENCES

1. Hill, M.C.; Tiedeman, C.R. *Effective Groundwater Model Calibration: With Analysis of Data, Sensitivities, Predictions, and Uncertainty*; John Wiley & Sons: New York, NY, USA, 2007; p. 480.
2. Carrera, J. An overview of uncertainties in modelling groundwater solute transport. *J. Contam. Hydrol.* **1993**, *13*, 23–48.

3. Kitanidis, P.K. *Introduction to Geostatistics: Applications in Hydrogeology*; Cambridge University Press: Cambridge, UK, 1997.

4. Yeh, W.W.-G. Review of parameter identification procedures in groundwater hydrology: The inverse problem. *Water Resour. Res.* **1986**, *22*, 95–108.

5. Ginn, T.R.; Cushman, J.H. Inverse methods for subsurface flow: A critical review of stochastic techniques. *Stoch. Hydrol. Hydraul.* **1990**, *4*, 1–26.

6. Gómez-Hernández, J.J.; Franssen, H.-J.H.; Sahuquillo, A. Stochastic conditional inverse modeling of subsurface mass transport: A brief review and the self-calibrating method.*Stoch. Environ. Res. Risk Assess.* **2003**, *17*, 319–328.

7. Hendricks Franssen, H.J.; Alcolea, A.; Riva, M.; Bakr, M.; van der Wiel, N.; Stauffer, F.; Guadagnini, A. A comparison of seven methods for the inverse modelling of groundwater flow. Application to the characterisation of well catchments. *Adv. Water Resour.* **2009**, *32*, 851–872.

8. Riva, M.; Guadagnini, A.; Neuman, S.P.; Janetti, E.B.; Malama, B. Inverse analysis of stochastic moment equations for transient flow in randomly heterogeneous media. *Adv. Water Resour.* **2009**, *32*, 1495–1507.

9. Carrera, J.; Alcolea, A.; Medina, A.; Hidalgo, J.; Slooten, L.J. Inverse problem in hydrogeology. *Hydrogeol. J.* **2005**, *13*, 206–222.

10. Bond, C.E.; Gibbs, A.D.; Shipton, Z.K.; Jones, S. What do you think this is? "Conceptual uncertainty" in geoscience interpretation. *GSA Today* **2007**, *17*, 4.

11. Neuman, S.P. Maximum likelihood Bayesian averaging of uncertain model predictions.*Stoch. Environ. Res. Risk Assess.* **2003**, *17*, 291–305.

12. Beven, K.; Binley, A. The future of distributed models: Model calibration and uncertainty prediction. *Hydrol. Process.* **1992**, *6*, 279–298.

13. Feyen, L.; Beven, K.J.; de Smedt, F.; Freer, J. Stochastic capture zone delineation within the generalized likelihood uncertainty estimation methodology: Conditioning on head observations. *Water Resour. Res.* **2001**, *37*, 625–638.

14. Freer, J.; Beven, K.; Ambroise, B. Bayesian estimation of uncertainty in runoff prediction and the value of data: An application of the GLUE approach. *Water Resour. Res.* **1996**, *32*, 2161–2173.

15. Hoeting, J.A.; Madigan, D.; Raftery, A.E.; Volinsky, C.T. Bayesian Model Averaging: A Tutorial. *Stat. Sci.* **1999**, *14*, 382–401.

16. Duan, Q.; Ajami, N.K.; Gao, X.; Sorooshian, S. Multi-model ensemble hydrologic prediction using Bayesian model averaging. *Adv. Water Resour.* **2007**, *30*, 1371–1386.

17. Rings, J.; Vrugt, J.A.; Schoups, G.; Huisman, J.A.; Vereecken, H. Bayesian model averaging using particle filtering and Gaussian mixture modeling: Theory, concepts, and simulation experiments. *Water Resour. Res.* **2012**, *48*, W05520.

18. Tsai, F.T.-C. Bayesian model averaging assessment on groundwater management under model structure uncertainty. *Stoch. Environ. Res. Risk Assess.* **2010**, *24*, 845–861.

19. Rojas, R.; Feyen, L.; Dassargues, A. Conceptual model uncertainty in groundwater modeling: Combining generalized likelihood uncertainty estimation and Bayesian model averaging. *Water Resour. Res.* **2008**, *44*, W12418.

20. Rojas, R.; Feyen, L.; Batelaan, O.; Dassargues, A. On the value of conditioning data to reduce conceptual model uncertainty in groundwater modeling. *Water Resour. Res.* **2010**, *46*, W08520.

21. Ye, M.; Neuman, S.P.; Meyer, P.D. Maximum likelihood Bayesian averaging of spatial variability models in unsaturated fractured tuff. *Water Resour. Res.* **2004**, *40*, W05113.

22. Ye, M.; Meyer, P.D.; Neuman, S.P. On model selection criteria in multimodel analysis. *Water Resour. Res.* **2008**, *44*.

23. Neuman, S.P.; Xue, L.; Ye, M.; Lu, D. Bayesian analysis of data-worth considering model and parameter uncertainties. *Adv. Water Resour.* **2012**, *36*, 75–85.

24. Xue, L.; Zhang, D.; Guadagnini, A.; Neuman, S.P. Multimodel Bayesian analysis of groundwater data worth. *Water Resour. Res.* **2014**, *50*, 8481–8496.

25. Chen, Y.; Zhang, D. Data assimilation for transient flow in geologic formations via ensemble Kalman filter. *Adv. Water Resour.* **2006**, *29*, 1107–1122.

26. Hendricks Franssen, H.J.; Kinzelbach, W. Real-time groundwater flow modeling with the Ensemble Kalman Filter: Joint estimation of states and parameters and the filter inbreeding problem. *Water Resour. Res.* **2008**, *44*.

27. Xie, X.; Zhang, D. Data assimilation for distributed hydrological catchment modeling via ensemble Kalman filter. *Adv. Water Resour.* **2010**, *33*, 678–690.

28. Xue, L.; Zhang, D. A multimodel data assimilation framework via the ensemble Kalman filter. *Water Resour. Res.* **2014**, *50*, 4197–4219.

29. Kurtz, W.; Hendricks Franssen, H.-J.; Brunner, P.; Vereecken, H. Is high-resolution inverse characterization of heterogeneous river bed hydraulic conductivities needed and possible?*Hydrol. Earth Syst. Sci.* **2013**, *17*, 3795–3813.

30. Burgers, G.; van Leeuwen, P.; Evensen, G. Analysis scheme in the ensemble Kalman filter.*Mon. Weather Rev.* **1998**, *126*, 1719–1724.

31. Samper, F.J.; Neuman, S.P. Estimation of spatial covariance structures by adjoint state maximum likelihood cross validation: 1. Theory. *Water Resour. Res.* **1989**, *25*, 351–362.

32. Deutsch, C.V.; Journel, A.G. *GSLIB: Geostatistical Software Library and User's Guide*, 2nd ed.; Oxford University Press Oxford: New York, NY, USA, 1998.

33. Ye, M.; Pohlmann, K.F.; Chapman, J.B.; Pohll, G.M.; Reeves, D.M. A model-averaging method for assessing groundwater conceptual model uncertainty. *Ground Water* **2010**, *48*, 716–728.

34. Poeter, E.; Anderson, D. Multimodel ranking and inference in ground water modeling.*Ground Water* **2005**, *43*, 597–605.

35. Banta, E.R. *MODFLOW-2000: The US Geological Survey Modular Ground-Water Model--Documentation of Packages for Simulating Evapotranspiration with a Segmented Function (ETS1) and Drains with Return Flow (DRT1), USGS Open-File Report 00–466*; US Geological Survey: Reston, VA, USA, 2000.

Chapter 11

Robust Non-Coherent Demodulation Scheme for Bluetooth Voice Transmission Using Linear, Extended, and Unscented Kalman Filtering

Ali S. Alghamdi[1,2,3], Mahdi N. Ali[4], Mohamed A. Zohdy[1]

[1]*Electrical and Computer Engineering Department, Oakland University, Rochester, MI, USA*

[2]*Saudi Red Crescent Authority, Riyadh, Saudi Arabia*

[3]*Ministry of Higher Education, Riyadh, Saudi Arabia*

[4]*Hyundai Motor Company, Superior Township, MI, USA*

ABSTRACT

This paper presents a novel and cost effective method to be used in the optimization of the Gaussian Frequency Shift Keying (GFSK) at the receiver of the Bluetooth communication system. The proposed method enhances the performance of the noncoherent demodulation schemes by improving the Bit Error Rate (BER) and Frame Error Rate (FER) outcomes. Linear, Extended, and Unscented Kalman Filters are utilized in this technique. A simulation model, using Simulink, has been created to simulate the Bluetooth voice transmission system with the integrated filters. Results have shown improvements in the BER and FER, and that the Unscented Kalman Filters (UKF) have shown superior performance in comparison to the linear Kalman Filter (KF) and the Extended Kalman Filter (EKF). To the best of our knowledge, this research is the first to propose the usage of the UKF in the optimization of the Bluetooth System receivers in the presence of additive white Gaussian noise (AWGN), as well as interferences.

Keywords: Bluetooth System Unscented Kalman Filter (UKF), Interferences, Gaussian Frequency Shift Keying (GFSK), Noncoherent Demodulation, Additive White Gaussian Noise (AWGN), Bit Error Rate (BER), Matlab Simulink

1. INTRODUCTION

In recent years, we have seen that the Bluetooth technology [1] becomes usable in diverse applications, in addition to being a standard feature in cell-phone devices. Due to its low costs, it has been used in most of the short distance communication systems, and there is a significant reason for that: the Bluetooth technology has free wireless bandwidth for short distance connection within 2.4 GHz frequencies.

The proposed method in this research is employed at the Gaussian Frequency Shift Keying (GFSK) modulation level in the physical layer of the Bluetooth system [2]. Bluetooth receivers implement on-coherent demodulation, which results in low cost products, but unreliable and poor performance in most applications. On the other hand, in order to optimize a non-coherent demodulation system, some complex Maximum Likelihood Sequence Estimators (MLSE) are used. Figure 1 shows a top level diagram of the Bluetooth system with our contribution of the robust Kalman Filter scheme. In the system block diagram, each block will be explained in detail in subsequent sections of this paper. We can see that the Kalman Filter (KF) is integrated in the receiver before the demodulation stage, so that it optimizes the detection processes. Furthermore, since we are dealing with a nonlinear system, the Extended Kalman Filter (EKF) and the Unscented Kalman Filter (UKF) are used in the optimization of the detected signals.

In 1960, Rudolf E. Kalman presented his idea of Kalman Filtering in [3]. KF is an estimator that estimates the states in linear dynamic systems by minimizing the Mean Square Error (MMSE). Kalman invented his filter based on modifying the Wiener-Kolmogorov filter, which was derived in 1940. While Wiener and Kolmogorov developed their filters based on frequency domain analysis, Kalman proposed his filter based on time domain analysis. The KF originated from Baye's probabilistic theory, in which, if a good number of past samples are known, then the future samples can be predicted and updated based on continuously collected results. Since most practical systems in real life are nonlinear, many research methods have been proposed to modify KF to be applied to nonlinear systems. The EKF has been proposed to be used in nonlinear systems by linearizing the states and measurement variables under study. In [4], the EKF has been used to optimize the received signal in the Bluetooth system in the presence of interferences, and produced better MSE in different Signal to Noise Ratio (SNR).

Literature [4] [5] has presented solid improvements, in which the KF proved to be useful in optimizing the Bluetooth system; however, the research in [5] investigated the performance of the system with the UKF. It concluded that the UKF produced a better outcome than the EKF. However, that paper performed the analysis without considering the interferences in the 2.4 GHz band in the AWGN. The work of this paper will continue on their work, which was listed in [4] [5]. In this research, the KF, EKF, and UKF are integrated into the Bluetooth receivers. The mathematical theory and derivation will be presented. The simulation model and results analysis will be analyzed and compared.

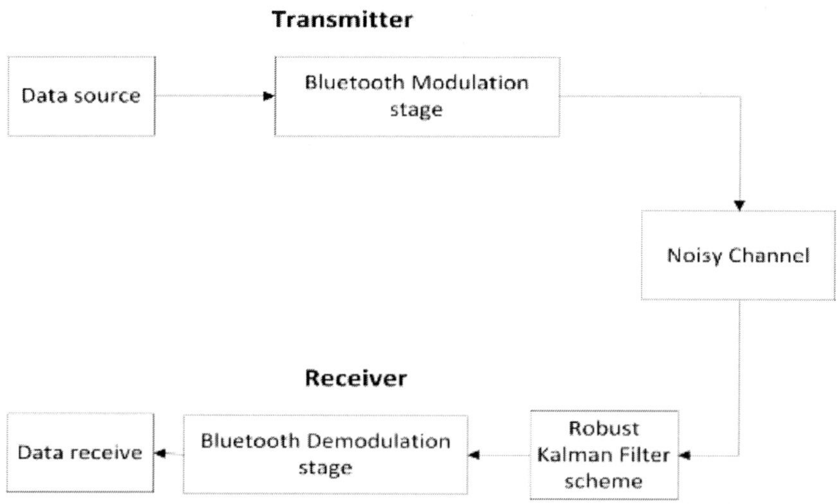

Figure 1. Bluetooth system with the integrated robust Kalman Filtering technique.

2. LITERATURE STUDY

There has been a great deal of literature, which has focused on the performance of the Bluetooth receivers. A significant amount of research has been performed to improve the non-coherent detection in the area of signal processing. Most of the research has developed techniques to improve the synchronization performance in low complexity demodulation schemes. The following sections describe the contributions that have been done in this related work.

Since 2003, up to the present, a GFSK has been optimized by many different techniques, which resulted in BER improvements. In [6], the authors investigated the performance of frequency hopped on GFSK for the Rayleigh fading channel; they showed an improvement in BER over the fading channel, which was corrupted by AWGN by the pilot assisted differential detection.

Meanwhile, the authors in [7] proposed a new digital receiver for a BT, where they combined a zero crossing demodulation with the de-correlating matched filter, and it also was tested on the BER performance versus h, the modulation index for GFSK.

The paper in [8] presents an improvement on the GFSK modulation method by using maximum likelihood sequence estimation (MLSE), based on a linearized model of the modulation. Furthermore, the authors were dedicated to evolving a scheme that had MLSE-oriented detection using the Viterbi equalizer considering a GFSK inherent inter symbol interference (ISI), with the channel gain sensitivity of 0 - 7 dB. In the article [9] , the author proposed a new receiver design, studying the sequence detection and discovering an algorithm, which they called non-coherent sequence detection (NSD). The new algorithm is unique in low complexity, as it operates distinctions caused by low-cost local oscillators. The NSD is combined with a forward error correction decoder, which yields optimal results in real Bluetooth devices. Another proceeding [10] developed a simulator for the Bluetooth scatternet, which had been tested in BER performance with an ad-hoc network using the ARQ scheme.

Some of these researches were proposed for the general purpose of optimization of signal detection for the GFSK modulation; hence, it can be considered for the Bluetooth systems improvement method. Digital carrier synchronization in [11] used a decision-aided data cancellation algorithm to compute the carrier frequency offset (CFO) for GFSK, where they measured the error difference due to the decision output between the I and Q receiver channels. The CFO will be taken into account, if the orthogonal frequency division multiplexing (OFDM) multi-user's technique is desirable, yet it is useful only for the multi-Piconet connection in a fading environment. Later on, a new optimization was published in a differential GFSK demodulator [12]. The author proposed the splitting of the transmitting data into a sequence of segments assigned different coefficients by setting up the first half of the sequence as linear and the other half as non-linear.

Some literature, which is the focus of our research, was published during the last four years. The article in [13] investigated the Bluetooth interference by enabling the Enhanced Data Rates (EDR) to run on three types of modulation techniques: Differential Quadratic Phase Shift Keying (π/4-DQPSK, 8-DPSK, and GFSK). Their result and analysis were impressive, because they applied to a real Bluetooth experimental application. The article produced significant results for the three types of modulations in the packet delay, which was stable and unaffected when the BER was 10^{-3} or higher. On the other hand, a study had been done on a similar topic in [14], in which the authors studied the interference of 802.11b WLAN, in which they developed a new Bluetooth voice packet synchronous connection oriented with a repeated transmission (SCORT) scheme. Their analysis focused on the interferences in the 2.4 GHz frequency unlicensed bandwidth, and their application was applied in a medium access control (MAC) layer. This paper focused on the future work as it is presented in [5], in which the authors proposed an optimization method of the GFSK and DPSK demodulation schemes using UKF, without considering the 802.11b interference factors in their work. The authors came up with an interactive Kalman Filtering technique to optimize the non-coherent GFSK and DPSK to produce better BER results. Finally, in [4] [15] -[17] , the authors made decent contributions using an EKF for GFSK, in order to estimate the phase of the signal at the receiver of the Bluetooth system, while considering the 802.11b interference factors in Gaussian noise and non-Gaussian noise environments.

3. GFSK SIGNAL DESCRIPTION

The GFSK modulation method is widely used in digital communications, including Bluetooth systems. Figure 2 illustrates the GFSK modulation system, and its modulated waveform can be represented as follows [3] [5] [18]:

$$S_m(t) = \mathrm{Re}\left[S_{ml}(t)e^{j2\pi f_c t}\right], \quad 1 \le m \le M, \ \ 0 \le t \le T \tag{1}$$

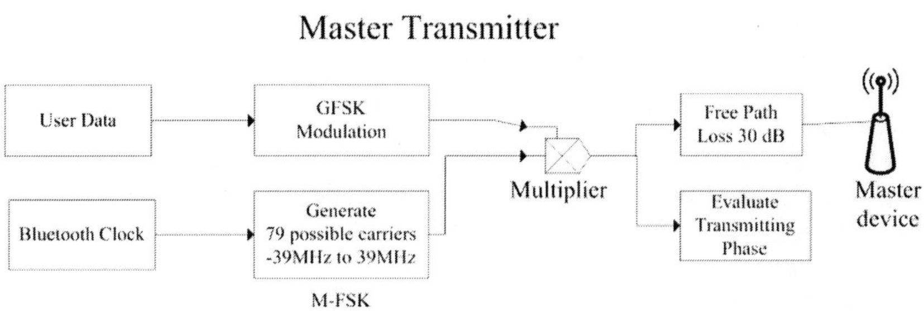

Figure 2. GFSK modulation block diagram.

$$= \sqrt{\frac{2E_s}{T_s}}\cos\left(2\pi f_c t + 2\pi m\Delta f t\right) \tag{2}$$

where

$$S_{ml}(t) = \sqrt{\frac{2E_s}{T_s}}e^{j2\pi m\Delta f t}, \quad 1 \le m \le M, \ \ 0 \le t \le T \tag{3}$$

Equation (2) is the formal GFSK signal generated by shifting the carrier within the amount of $m\Delta f$, $1 \le m \le M$, to reflect the digital information that will be transmitted. This GFSK modulation is different from the Frequency Shift Keying (FSK), because we need to avoid using signals having large spectral side lobes. The information-bearing signal frequency modulates a signal carrier whose frequency is changed continuously. This is the result of the modulated signal, which is phase-continuous, and is called the Continuous-Phase Frequency Shift Keying (CPFSK); hence, this type of signal has memory, and the phase of the carrier is constrained to be continuous [18].

A complete presentation of CPFSK will start to drive the signal representation from the Pulse Amplitude Modulation (PAM):

$$d(t) = \sum_{n} I_n g(t - nT_s) \tag{4}$$

where $\{I_n\}$ denoted the sequence of the amplitudes obtained by mapping k-bit block in M-array, information symbols $\{a_n\}$ can be represented in the format of $\pm 1, \pm 2, \pm 3, \cdots, \pm (M-1)$, and $g(t)$ is a rectangular pulse of the amplitude $\frac{1}{2T_s}$, and duration T_s seconds. The signal $d(t)$ is used to modulate the frequency of the carrier. Consequently, the complex baseband waveform $v(t)$ can be expressed as:

$$v(t) = \sqrt{\frac{2E_s}{T_s}} e^{j\left[4\pi T_s f_d \int_{-\infty}^{t} d(\tau) d\tau + \phi_0\right]} \tag{5}$$

where f_d is the peak frequency deviation and ϕ_0 is the initial phase of the carrier. Therefore, by using Euler's formula in order to extend an exponential complex function Equation (5) can be expressed as:

$$s(t)=\sqrt{\frac{2E_s}{T_s}}\cos\left[2\pi f_c t+\phi(t;\boldsymbol{I})+\phi_0\right] \tag{6}$$

The $\phi(t;\boldsymbol{I})$ is the time-varying phase of the carrier, which is distinct as:

$$\phi(t;\boldsymbol{I})=4\pi T_s f_d\int_{-\infty}^{t}\mathrm{d}(\tau)\,\mathrm{d}\tau \tag{7}$$

$$=4\pi T_s f_d\int_{-\infty}^{t}\left[\sum_n I_n g(\tau-nT)\right]\mathrm{d}\tau \tag{8}$$

Carrying out the integral of the phase carrier in the interval $nT_s\le t\le(n+1)T_s$ can be represented in form:

$$.\ \phi(t;\boldsymbol{I})=2\pi f_d T_s\sum_{k=-\infty}^{n-1}I_k+2\pi f_d q(t-nT_s)I_n \tag{9}$$

$$=\theta_n+2\pi h I_n q(t-nT_s) \tag{10}$$

where:

$$h=2f_d T_s \tag{11}$$

$$\theta_n=\pi h\sum_{k=-\infty}^{n-1}I_k \tag{12}$$

$$q(t)=\begin{cases}0, & t<0\\ \frac{t}{2T_s}, & 0\le t\le T_s\\ \frac{1}{2}, & t>0\end{cases} \tag{13}$$

The parameter h is called the modulation index. However, we show the mathematical GFSK modulation proof, which has already been discussed in [18]. The GFSK has the property of a Gaussian filter applied to a square signal waveform. Since the square signal has either 0 s or 1 s for output, which represent bits, and when we multiply the Gaussian filter, the frequency increases when it has symbol 1, and decreases when the symbol is 0, which all describes the frequency modulation (FM) criteria. The Gaussian filter can be described as follows:

$$x(t)=\frac{1}{\sqrt{2\pi}\sigma}\mathrm{e}^{-\frac{1}{2}\left(\frac{t}{\sigma}\right)^2} \tag{14}$$

$$\sigma=\frac{\sqrt{\ln 2}}{2\pi A} \tag{15}$$

where σ is related to the filters 3 dB bandwidth A.

4. NOISY CHANNEL

The employed channel is an additive white Gaussian noise (AWGN), which is widely and extensively used in wireless communication, especially with short distance connections [9]. The receiving signal can be represented as:

$$r(t)=\mathrm{e}^{j\phi(t)}s(t)+I(t)+n(t) \tag{16}$$

where $s(t)$ is described in Equation (6), $I(t)$ represents the 802.11b interference with 40 dB free space loss, and $n(t)$ is the white Gaussian noise AWGN carrying $10\ \text{mW}$ noise power.Figure 3 describes the channel structure.

802.11 Interference Model

IEEE 802.11b is an exponential model developed for a 2.4 GHz indoor channel. It is used for small coverage areas, such as a Wi-Fi network inside a building. However, the channel conditions may vary with time and location, due to the environment being enclosed by walls, which still requires a power delay profile (PDP) to characterize their channel delay and power average. The channel impulse response can be calculated through the output of a finite impulse response (FIR) filter; of course, each channel is modeled by a complex Gaussian random variable with its average power. In order to determine the maximum number of paths, we need to calculate the RMS delay σ_τ, and sampling period T_s, as [23]:

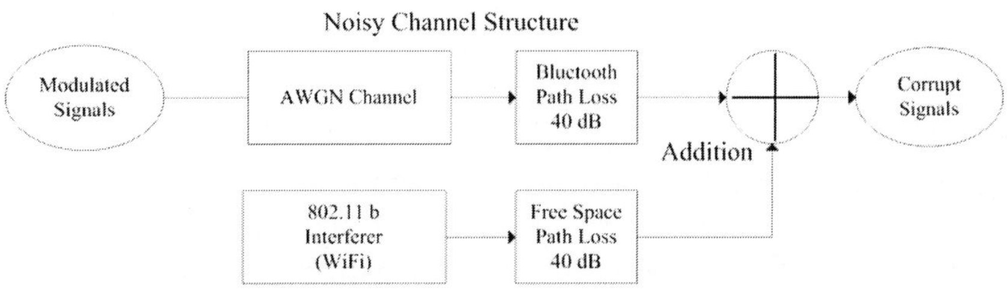

Figure 3. Noisy channel structure.

$$p_{\max} = 10\frac{\sigma_\tau}{T_s} \tag{17}$$

Let us consider that we have a Gaussian distribution $\mathcal{N}\left(0, \frac{\sigma_p^2}{2}\right)$, with mean and variance respectively. The maximum excess delay, according to the specification of the 802.11 channels, is 10 times the RMS delay. So, the power of each channel can be represented as:

$$\sigma_p^2 = \sigma_0^2 \mathrm{e}^{\frac{pT_s}{\sigma_\tau}} \tag{18}$$

where σ_0^2 is:

$$\sigma_0^2 = \frac{1-\mathrm{e}^{\frac{-T_s}{\sigma_\tau}}}{1-\mathrm{e}^{\frac{-(p_{\max}+1)T_s}{\sigma_\tau}}} \tag{19}$$

Figure 4 is an illustration of an example of the 802.11 indoor channel impulse response, and its power average, when $\sigma_\tau = 20$ ns, and $T_s = 50$ ns.

5. NON-COHERENT DETECTION

The detection is part of the demodulation stage at the receiver side of the BT in Figure 1. In BT, we do not have a reliable receiver, or, in other words, the transmitter does not have a distinguished receiver, which has none of the transmitting signal property. Therefore, they are called a non-coherent detection, due to the invalid assumption of the signals $\{S_m(t), 1 \le m \le M\}$, or the availability of an orthonormal basis $\{\phi_j(t), 1 \le j \le N\}$ [18] , caused by either a random attenuation, or a random phase shift impact, of the signals. Hence, the non-coherent GFSK is modulated in the identity, when the signal phase matter for detection,Figure 5, showing how the phase trajectory for each bit can be acknowledged [18] .

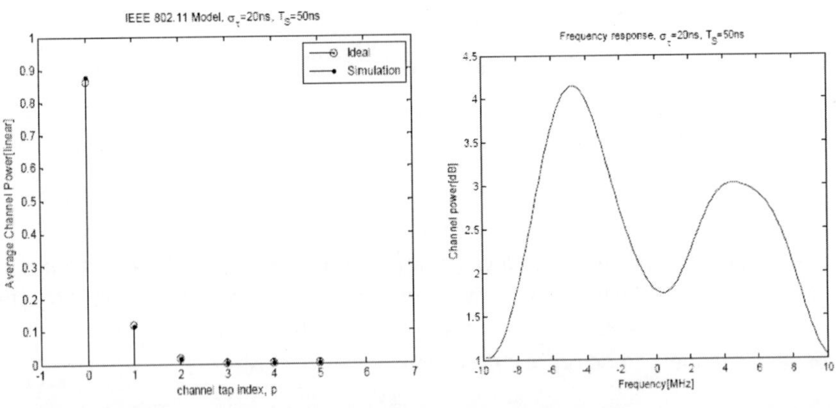

Figure 4. 802.11b channel impulse example with their power.

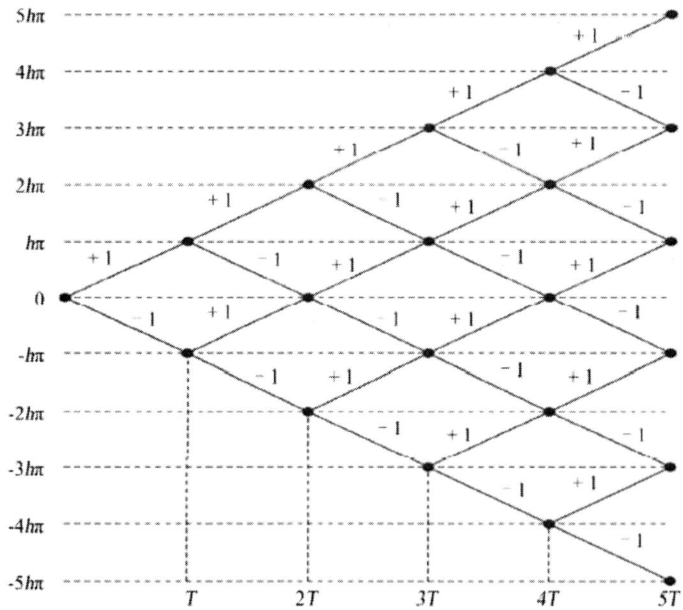

Figure 5. Phase trajectory for binary GFSK.

Obviously, Figure 5 provides a proof of our statement, when we address the GFSK modulation as a product of the periodic square signal and the Gaussian filter, so the phase increases $+A$ (amplitude) when (binary 1) is transmitted, and decreases $-A$ when (binary 0) is transmitted. The optimal non-coherent detection of GFSK is the real convolution of the complex baseband waveform Equation (3):

$$\mathrm{Re}\left[\int_{-\infty}^{\infty} S_{ml}(t) S_{m'l}(t)\mathrm{d}t\right] = 2\varepsilon_s \mathrm{sinc}\left[2(m-m')\Delta f T_s\right] \tag{20}$$

where ε_s is denoted as the energy per transmitted symbol, Δf is the frequency separation between signals, and T_s is the pulse period. Therefore, the general BER for GFSK can be evaluated using the general form of BER for any digital communication [2].

$$P_e = P(S_1\text{sent})\int_{-\infty}^{v_T} f(\mu_0|S_1)\mathrm{d}\mu_0 + P(S_2\text{sent})\int_{v_T}^{\infty} f(\mu_0|S_2)\mathrm{d}\mu_0 \tag{21}$$

In this case v_T is the threshold voltage, and μ_0 is the test statistic, which is a basic random variable. Then,

with the probability of transmitting S_1, and $S_2 = \frac{1}{2}$, and an optimum threshold $v_T = 0$, Equation (21) yields:

$$P_e = \frac{1}{2} e^{-\frac{\varepsilon_b}{2N_0}} \tag{22}$$

6. GFSK STATE SPACE MODEL

A GFSK module signal needs to be modified according to the KF procedure, because KF operates in two steps: a prediction and an update, so GFSK signals have a magnitude and a phase angle. For non-coherent detection, we proposed the phase estimation, in order to detect the information using the phase trajectory illustrated in Figure 5. Prediction and update processes have the relation between the previous phase and current phase estimations, finding the error among them, and then predicting the next phase.

GFSK Description

In the KF development procedure, the implementation took place to evaluate or estimate in two steps: predict and update. The prediction has been computed by KF to evaluate the current phase and the previous state phase, finding the error, to update the future state, using the residual calculated error. GFSKs have a magnitude and a phase angle, which are unpredictable, due to the noise effect. Our KF are involved in two states to estimate the phase $\theta(t;\boldsymbol{I}_n)$, and the difference equation $\Delta\phi(t;\boldsymbol{I}_n)$, where $\boldsymbol{I}_n$ input data is a modulated phase [5]. The dynamic system model is represented as follows:

$$\theta(t;\boldsymbol{I}_n)_k = \theta(t;\boldsymbol{I}_n)_{k-1} + h f\left(\theta(t;\boldsymbol{I}_n)\right)_{k-1}, \quad kT_s \le t \le (k+1)T_s \tag{23}$$

$$\left(\Delta\phi(t;\boldsymbol{I}_n)_{k-1}\right)^2 = \theta(t;\boldsymbol{I}_n)_k - \theta(t;\boldsymbol{I}_n)_{k-1} + W_k, \quad kT_s \le t \le (k+1)T_s \tag{24}$$

$\text{Hint!}: W_k = I(t) + n(t)\cdots$ in Equation (16) (25)

Now, our desirable result is to estimate the phase $\theta(t;I_n)_k$, and that can be computed using Euler's identity for complex exponential functions. Therefore, the difference in the phase $hf(\theta(t;I_n))_{k-1}$ is evaluated by summing the current phase $\theta(t;I_n)_k$ with the previous change in the phase, where h is the GFSK modulation index, and W_k is defined in Equation (24) summation of AWGN and 802.11b interference noise.

$$\theta(t;I_n)_k = \theta(t;I_n)_{k-1} + \left(\Delta\phi(t;I_n)_{k-1}\right)^2 \quad (26)$$

Equation (23) shows that the system begins the module in nonlinear format, and the receiving signal can be written as:

$$r(t) = \sqrt{\frac{2E_s}{T_s}} e^{j\left[\theta(t;I_n)_{k-1} + \left(\Delta\phi(t;I_n)_{k-1}\right)^2\right]} + v(t) \quad \text{for } kT_s \le t \le (k+1)T_s \quad (27)$$

7. STATE MODEL

By substituting Equation (26) into Equations (23) and (24), the state difference equations can be formalized as:

$$\begin{aligned} \theta(t;I_n)_k &= \theta(t;I_n)_{k-1} + \left(\Delta\phi(t;I_n)_{k-1}\right)^2 \\ \Delta\phi(t;I_n)_k &= \Delta\phi(t;I_n)_{k-1} + W_k \end{aligned} \quad (28)$$

Noise Measurement Model

According to the non-coherent detection, the typical receiver of AWGN has a correlated process when the receiving signal is divided in two phases: in-phase and quadratic phase components, which are the more familiar phase I and Q receiver channel when $t = k$.

$$Z_k = h(\theta_k) + v_k = \sqrt{\frac{2E_s}{T_s}} \begin{bmatrix} \cos\left(\theta(t; I_n)_k\right) \\ \sin\left(\theta(t; I_n)_k\right) \end{bmatrix} + \begin{bmatrix} v_{I_k} \\ v_{Q_k} \end{bmatrix} \tag{29}$$

8. PROPOSED SCHEME ANALYSIS

Referring to Figure 1, the first stage at the receiver, which we have called the "Robust Kalman Filter Scheme", yields our contributed scheme, yet we are using a KF, which is one of the adaptive filters for signal processing manner. Although in this paper we adjust our results in three types of filters: linear KF, EKF, and UKF, all of them have been extensively used in stochastic random signal processing. The result could be a rich optimization for a wireless communication connection. Moreover, the author in [5] implemented an interactive KF, using a two-modulation technique, and simultaneously concluded his article by proving the UKF had a better result, but his effort was in the system with no interference, assuming the BT system was operated in an area such as a vehicle, where there is not much 802.11b interference. Nonetheless, the authors in [4] [15] -[17] have also contributed meaningfully to the same topic, to improve the non-coherent detection for GFSK proposed in their work for the BT system, by involving 802.11b interference in their results. Hence, their efforts were published, concluding with using the EKF for the purpose of a nonlinear optimization minimizing the mean square error (MES).

Beginning with our scheme description, we are designing three types of stochastic filters, as we mentioned earlier: linear KF, EKF, and UKF, while we have fully considered the latest literature review, as applied to the BT voice transmission system [19] . Consequently, we will be presenting our schemes analysis in three sections, each one discussing the details of our filters, and Figure 6 gives an image of our robust scheme that cascaded with the demodulation stage.

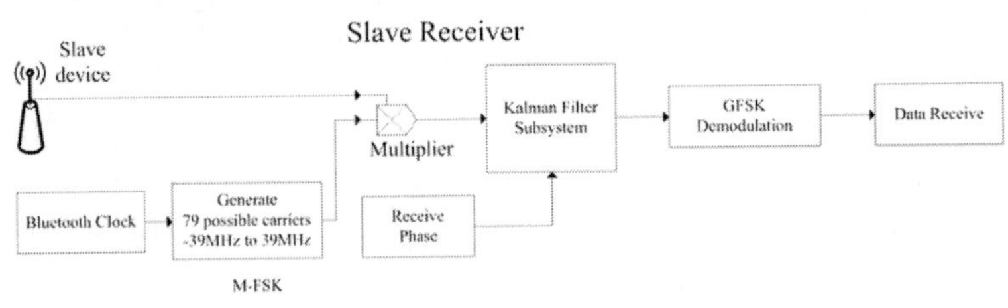

Figure 6. Slave non-coherent demodulation black diagram.

In the early 1960s, R. E. Kalman considered the same problem that Wiener had dealt with. In his 1960 paper, he proposed the noise measurement to be a discrete sequence in time, applied to a continuous-time signal [20] . In addition, the KF recursive equation was treated as a computational loop between an input and an output of the dynamic system, so the loop had two major tasks: predict and update. The following is the general form of the linear KF equations [20] [21]:

$$\boldsymbol{X}[k+1] = Ax[k] + Gw[k] \tag{30}$$

$$\boldsymbol{Z}[k] = Cx[k] + v[k] \tag{31}$$

where:

$\boldsymbol{X}_k$ = process state vector at time t_k.

A_k = State Transition matrix.

G_k = noise coefficient.

$\boldsymbol{Z}_k$ = vector measurment at time t_k.

C_k = ideal noiseless connection between the measurementand the state vector at time t_k.

v_k = measurement error.

Now, for our system, the general Kalman state equations need to modify the base on the signal representation of GFSK, so we let $\boldsymbol{X}_k = \theta_k$. We assume the initial estimation of some point in time, t_k, and that is based on our knowledge of the process prior to t_k. In the process of prediction, the phase θ_k takes the difference from the previous phase angle, and now we can drive from the Equation (24):

$$(\Delta\phi_k)^2 = \theta_k - \theta_{k-1} \tag{32}$$

Nonetheless, we need to have $(\Delta\phi_k)^2$, so then we can compute θ_k, using the joint estimation property $\theta_k = \theta_{k|k-1}$; the Equation (32) can be written as:

$$(\Delta\phi_k)^2 = \theta_{k|k-1} - \theta_{k-1} \tag{33}$$

The prior (or a priori) estimate will be denoted as $\hat{\theta}_k^-$ where the $\hat{\theta}_k$ denotes the estimate, and $\hat{\theta}_k^-$, with "super minus", is denoted as our best estimate prior to assimilating the measurement at t_k. We can define the estimation error to be:

$$e_k^- = \theta_k - \hat{\theta}_k^- \tag{34}$$

8.1. Linear Kalman Filter (KF)

First of all, we have to rewrite a KF state space model, corresponding to our GFSK modulated signal, carried in the general KF format in Equations (30) and (31).

$$\hat{\theta}_k = A_k\hat{\theta}_{k-1} + G_kW_k \tag{35}$$

$$Z_k = C_k\theta_k + V_k \tag{36}$$

For linear estimation, we have two types of covariance matrices that need to be considered for W_k and V_k vectors, which are given as [20]:

$$E\left[W_kW_i^{\mathrm{T}}\right] = \begin{cases} Q_k, & i = k \\ 0, & i \neq k \end{cases} \tag{37}$$

$$E\left[V_k V_i^{\mathrm{T}}\right]=\begin{cases}R_k, & i=k\\0, & i\neq k\end{cases} \tag{38}$$

$$E\left[W_k V_i^{\mathrm{T}}\right]=0,\ \text{for all } k \text{ and } i \tag{39}$$

Corresponding to error Equation (34), the error covariance matrix is:

$$P_k^- = E\left[e_k^- e_k^{-\mathrm{T}}\right] = E\left[\left(\theta_k-\hat{\theta}_k^-\right)\left(\theta_k-\hat{\theta}_k^-\right)^{\mathrm{T}}\right] \tag{40}$$

Now, we are looking to use the measurement Z_k to improve the prior estimate, so we assume a linear blending of the noisy measurement and the prior estimation in the equation:

$$\hat{\theta}_k = \hat{\theta}_k^- + K_k\left(Z_k - C_k\hat{\theta}_k^-\right) \tag{41}$$

where $\hat{\theta}_k$ is the update estimation, and K_k is the blending coefficient. Thus, by using the minimum mean square error (MMSE), we can find an optimal K_k. In order to determine the MMSE, we need to handle the expression of the error covariance in matrix form in relation to (a posteriori) the estimate:

$$P_k = E\left[e_k e_k^{\mathrm{T}}\right] = E\left[\left(\theta_k-\hat{\theta}_k\right)\left(\theta_k-\hat{\theta}_k\right)^{\mathrm{T}}\right] \tag{42}$$

By substituting the Equation (36) into Equation (41), we will yield a result that has been substituted in Equation (42), and illustrate that $\left(\theta_k-\hat{\theta}_k^-\right)$ is the a priori estimation error, which is uncorrelated with measurement error . Although, from the algebraic substitutions, we will be end up with a general expression for the update er- ror covariance matrix.

$$\boldsymbol{P}_k = \left(\boldsymbol{I} - \boldsymbol{K}_k \boldsymbol{C}_k\right) \boldsymbol{P}_k^- \left(\boldsymbol{I} - \boldsymbol{K}_k \boldsymbol{C}_k\right)^{\mathrm{T}} + \boldsymbol{K}_k \boldsymbol{R}_k \boldsymbol{K}_k^{\mathrm{T}} \tag{43}$$

Expanding Equation (43) becomes:

$$\boldsymbol{P}_k = \boldsymbol{P}_k^- - \boldsymbol{K}_k \boldsymbol{C}_k \boldsymbol{P}_k^- - \boldsymbol{P}_k^- \boldsymbol{C}_k^{\mathrm{T}} \boldsymbol{K}_k^{\mathrm{T}} + \boldsymbol{K}_k \left(\boldsymbol{C}_k \boldsymbol{P}_k^- \boldsymbol{C}_k^{\mathrm{T}} + \boldsymbol{R}_k\right) \boldsymbol{K}_k^{\mathrm{T}} \tag{44}$$

Optimizing $\boldsymbol{P}_k$ by process, two differentiations produce the trace of $\boldsymbol{P}_k$ with respect to $\boldsymbol{K}_k$, then solve for $\boldsymbol{K}_k$:

$$\boldsymbol{K}_k = \boldsymbol{P}_k^- \boldsymbol{C}_k^{\mathrm{T}} \left(\boldsymbol{C}_k \boldsymbol{P}_k^- \boldsymbol{C}_k^{\mathrm{T}} + \boldsymbol{R}_k\right)^{-1} \tag{45}$$

Thus, the Equation (45) gives the suitable value of $\boldsymbol{K}_k$, which, in a sense, will control minimizing the MSE estimation, and is called Kalman gain. Finalizing the computation of $\boldsymbol{P}_k$ by having a routine substitution Equation (45) in Equation (44) leads to:

$$\boldsymbol{P}_k = \boldsymbol{P}_k^- - \boldsymbol{K}_k \left(\boldsymbol{C}_k \boldsymbol{P}_k^- \boldsymbol{C}_k^{\mathrm{T}} + \boldsymbol{R}_k\right) \boldsymbol{K}_k^{\mathrm{T}} = \left(\boldsymbol{I} - \boldsymbol{K}_k \boldsymbol{C}_k\right) \boldsymbol{P}_k^- \tag{46}$$

Hence, we still need to find $\hat{\theta}_k^-$ and $\boldsymbol{P}_k^-$, which are our desirable best estimations; this has to be done in a similar process, due to the use of the measurement Z_{k+1} in Equation (36). From Equation (35), neglecting the term $G_k W_k$, since we have the AWGN with zero mean, and, of course, not correlated, so we have:

$$\hat{\theta}_{k+1}^- = A_k \hat{\theta}_k \tag{47}$$

Back to using the error estimation Equation (34), according to Equation (44), we have:

$$e_{k+1}^- = \theta_{k+1} - \hat{\theta}_{k+1}^- = A_k e_k + w_k \tag{48}$$

$$P_{k+1}^{-} = E\left[e_{k+1}^{-}e_{k+1}^{-\mathrm{T}}\right] = E\left[\left(A_k e_k + w_k\right)\left(A_k e_k + w_k\right)^{\mathrm{T}}\right] = A_k P_k A_k^{\mathrm{T}} + Q_k \tag{49}$$

So far, we derived the entire KF algorithm processes associated with the modulated BT system, and Figure 7 shows the block diagram of the KF algorithm to prove the statement of recursive equation.

8.2. The Extended Kalman Filter (EKF)

It was mentioned earlier, in Section 2, that [4] [15] -[17] have made meaningful contributions in the same subject that with which we are working, especially in using EKF for a non-coherent BT receiver. Furthermore, proposing an EKF for this article becomes more a matter of comparison, and some of the better achievements will be illustrated in Section 10. The EKF has been developed for nonlinear dynamic systems, and nonlinear measurement relationships, certainly, will require a linearization process for prediction and update measurements. Thus, the linearization will accommodate the general linear KF system module, so, starting with our system, we have an associated measurement relationship that may be written as [20]:

$$\dot{\theta}_k = f\left(\theta_k, u_d, t\right) + W_k\left(t\right) \tag{50}$$

$$\mathbf{Z}_k = h\left(\theta_k, t\right) + V_k\left(t\right) \tag{51}$$

where f and h are known functions, u_d is a forcing function, and v and w are uncorrelated white noise processes. For linearization, we may choose $\theta_k^*(t)$ to be an approximate trajectory determined by the mean.

$$\theta_k = \theta_k^*\left(t\right) + \Delta\theta\left(t\right) \tag{52}$$

So, Equations (50) and (51) can be written as:

$$\dot{\theta}_k^* + \Delta\dot{\theta} = f\left(\theta_k^* + \Delta\theta, u_d, t\right) + W_k(t) \tag{53}$$

$$\mathbf{Z}_k = h\left(\theta_k^* + \Delta\theta, t\right) + V_k(t) \tag{54}$$

Now, we approximate the f and h functions by using Tylor's series expansions:

$$\theta_k^* + \Delta\dot{\theta} \approx f\left(\theta_k^*, u_d, t\right) + \left[\frac{\partial f}{\partial \theta}\right]_{\theta_k = \theta_k^*} \Delta\theta + W_k(t) \tag{55}$$

$$\mathbf{Z}_k \approx h\left(\theta_k^*, t\right) + \left[\frac{\partial h}{\partial \theta}\right]_{\theta_k = \theta_k^*} \Delta\theta + V_k(t) \tag{56}$$

$$\frac{\partial f}{\partial \theta} = \begin{bmatrix} \frac{\partial f_1}{\partial \theta_1} & \cdots & \frac{\partial f_1}{\partial \theta_k} \\ \vdots & \ddots & \vdots \\ \frac{\partial f_k}{\partial \theta_1} & \cdots & \frac{\partial f_k}{\partial \theta_k} \end{bmatrix}; \quad \frac{\partial h}{\partial \theta} = \begin{bmatrix} \frac{\partial h_1}{\partial \theta_1} & \cdots & \frac{\partial h_1}{\partial \theta_k} \\ \vdots & \ddots & \vdots \\ \frac{\partial h_k}{\partial \theta_1} & \cdots & \frac{\partial h_k}{\partial \theta_k} \end{bmatrix}, \quad k = 1, 2 \cdots \tag{57}$$

Equation (57) is the Jacobian matrix, and it allows having the nominal trajectory of $\theta_k^*(t)$ to satisfy the deterministic differential equation. From that, we can give the linearized equation as:

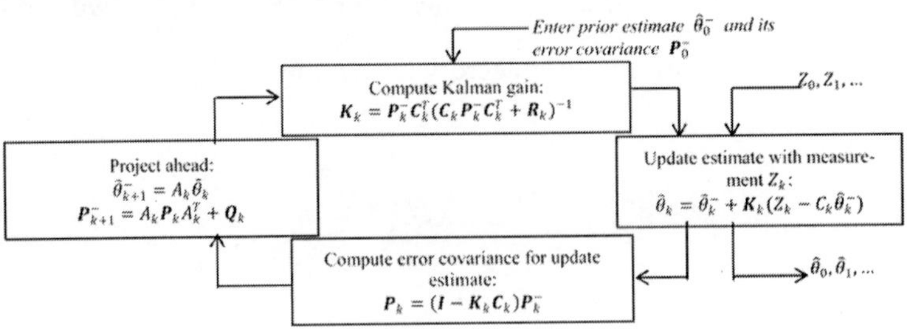

Figure 7. Kalman recursive algorithm.

$$\Delta\dot{\theta} = \left[\frac{\partial f}{\partial \theta}\right]_{\theta_k = \theta_k^*} \Delta\theta + W_k(t) \quad \text{(linearized dynamics)} \tag{58}$$

$$\left[z - h\left(\theta_k^*, t\right)\right] = \left[\frac{\partial h}{\partial \theta}\right]_{\theta_k = \theta_k^*} \Delta\theta + V_k(t) \quad \text{(linearized measurement equation)} \tag{59}$$

Finally, the EKF can apply to the linear KF algorithm shown in Figure 7; once the linearization process is taken care of, the sequences and the coefficients of the measurements (A_k, C_k) can be evaluated as the following:

$$A_k = \left.\frac{\partial A_k}{\partial \theta}\right|_{\theta_{k-1}^*} = \begin{bmatrix} 1 & 1 \\ 0 & 1 \end{bmatrix} \tag{60}$$

$$C_k = \left.\frac{\partial C_k}{\partial \theta}\right|_{\theta_k^*} = \begin{bmatrix} -g\left(\text{sing}^{\mathrm{T}}\theta_k\right) & 0 \\ g\left(\text{cosg}^{\mathrm{T}}\theta_k\right) & 0 \end{bmatrix}, \quad g = \begin{bmatrix} 1 \\ 0 \end{bmatrix} \tag{61}$$

8.3. The Unscented Kalman Filter (UKF)

For random variables (RV), or Gaussian random variables (GRV), the UKF proposed an extension development of EKF, in particular, for nonlinear statistical estimation regarding the main principles of linear KF, which are prediction and update. Obviously, the name of UKF came after the unscented transformation (UT), which developed the solution; in another words, the UT evaluates the mean and covariance (μ_x and C_x) respectively, of the RV, based on discrete samples projected through the associated nonlinear transformation [20] . Furthermore, the UT carries the mean and covariance of the data with the last two information samples, where these samples related to each other on the standard deviation and approximate dimension; this relation creates the sigma points from the mean, in Figure 8, showing the UT [21] [22] .

Assuming a nonlinear function $f(.)$ that propagates an RV x , to yield a result in another RV $y = f(x)$, so let us derive the UT from the sample point to estimate the mean and covariance in y (μ_y and C_y) respectively.

(i) Scaling the unscented samples:

$$\mathfrak{X}_i = \begin{cases} (\mu_x)_i, & i = 0 \\ \left(\mu_x + \sqrt{(N+\lambda)C_x}\right)_i, & i = 1,\cdots,N \\ \left(\mu_x - \sqrt{(N+\lambda)C_x}\right)_i, & i = N+1,\cdots,2N \end{cases} \tag{62}$$

where:

$$\lambda = \alpha^2 (N + k) - N$$

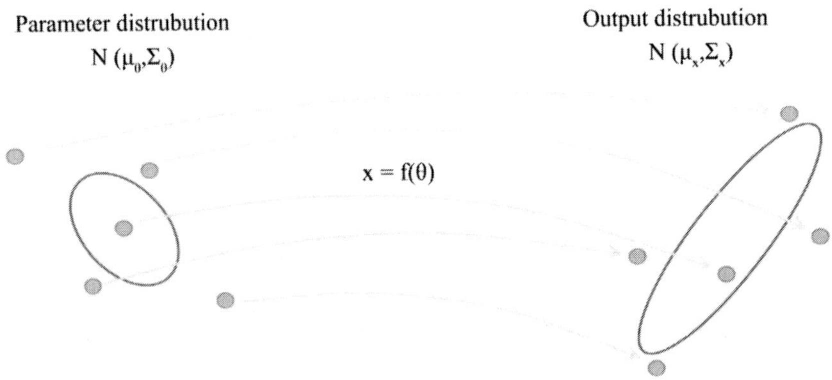

Figure 8. The principle of UT.

N = dimension of state x

α = determines spread of the unscented samples

β = dependent on knowledge of distrubution

k = scaling factor

(ii) Transforming the unscented samples: $\mathcal{Y}_i = f(\mathfrak{X})$.

(iii)

$$\text{Samples weights} = \begin{cases} \omega_0^{\mu} = \dfrac{\lambda}{\lambda + N} \\ \omega_0^{\mu} < i \le 2N = \dfrac{1}{2(\lambda + N)} \\ \omega_0^{C} = \omega_0^{\mu} + 1 - \alpha^2 + \beta \\ \omega_0^{C} < i \le 2N = \dfrac{1}{2(\lambda + N)} \end{cases} \quad (63)$$

(iiii)

$$\hat{\mu}_y = \sum_{i=0}^{2N} \omega_i^C \mathcal{Y}_i \quad (64)$$

$$C_y = \sum_{i=0}^{2N} \omega_i^C \left(\mathcal{Y}_i - \hat{\mu}_y\right)\left(\mathcal{Y}_i - \hat{\mu}_y\right)^{\mathrm{T}} \quad (65)$$

Equations (64) and (65) describe how UT computes the mean and covariance; also, Figure 9visualizes the UT data process. Now, it is necessary to run the linear KF through the output of UT, starting with the basic form of the linear KF dynamic system, involving the process update and measurement update.

The UKF algorithm starts from the original linear KF measurement model, and then takes inside an initial state where $\theta = \mathfrak{X}$ for the UT processes as follows:

$$\theta_{k+1} = f\left(\theta_k\right) + W_k \quad (66)$$

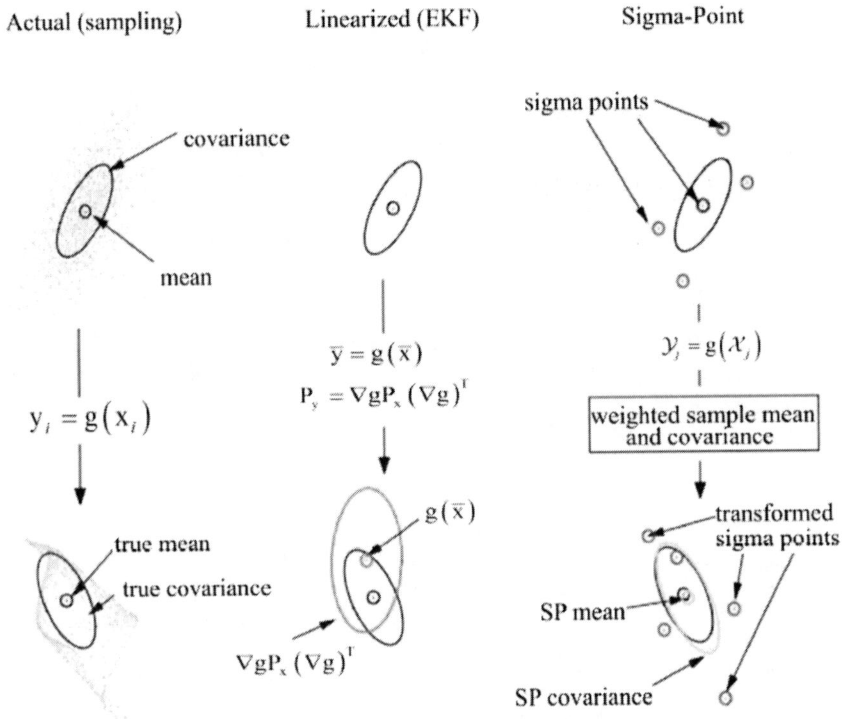

Figure 9. UT mean and covariance processes.

$$z_k = h(\theta_k) + V_k \tag{67}$$

Equations (66) and (67) are the nonlinear UKF equations with noise inputs W_k and V_k, respectively for the BT system, which will be considered white Gaussian noise [20] [21] .

Initialize:

$$\hat{\theta}_0 = E[\theta_0] \tag{68}$$

$$P_0 = E\left[(\theta_0 - \hat{\theta}_0)(\theta_0 - \hat{\theta}_0)^T\right] \tag{69}$$

Calculating the sigma points:

$$\mathfrak{X}_{k-1} = \left[\hat{\theta}_{k-1}\hat{\theta}_{k-1} + \sqrt{(N+\lambda)P_{k-1}}\hat{\theta}_{k-1} - \sqrt{(N+\lambda)P_{k-1}}\right] \tag{70}$$

Time update processes:

The prediction of UKF projects the set of samples through the nonlinear Equation (66) as $\mathfrak{X}_k^- = f(\mathfrak{X}_{k-1})$, and then reconstitutes the a priori state estimation. For the record, as mentioned earlier, a "super minus" script means the optimal, or the best, estimation.

$$\hat{\theta}_k^- = \sum_{i=0}^{2N} \omega_i^{\mu} \mathfrak{X}_k^{(i)-} \tag{71}$$

$$P_k^- = \left\{\sum_{i=0}^{2N} \omega_i^C \left(\mathfrak{X}_k^{(i)-} - \hat{\theta}_k^-\right)\left(\mathfrak{X}_k^{(i)-} - \hat{\theta}_k^-\right)^{\mathrm{T}}\right\} + \boldsymbol{Q}_k \tag{72}$$

Thus, after computing the optimal mean Equation (71), and its error covariance Equation (72), we can re- sample a new sigma point for the remainder from the following relationship:

$$\mathfrak{X}_k^- = \hat{\theta}_k^- \hat{\theta}_k^- + \sqrt{(N+\lambda)P_k^-}\hat{\theta}_k^- - \sqrt{(N+\lambda)P_k^-} \tag{73}$$

The next steps are the measurement update processes, which required the sigma points operated through the nonlinear Equation (67).

$$\mathfrak{X}_k^- = h\left(\mathfrak{X}_k^{(i)-}\right) \tag{74}$$

$$\hat{z}_k^- = \sum_{i=0}^{2N} \omega_i^{\mu} \mathfrak{X}_k^{(i)-} \tag{75}$$

$$\left(P_{\hat{z}\hat{z}}^-\right)_k = \left\{\sum_{i=0}^{2N} \omega_i^C \left(\mathfrak{X}_k^{(i)-} - \hat{z}_k^-\right)\left(\mathfrak{X}_k^{(i)-} - \hat{z}_k^-\right)^{\mathrm{T}}\right\} + \boldsymbol{R}_k \tag{76}$$

$$\left(P_{\hat{\theta}\hat{z}}^{-}\right)_k = \sum_{i=0}^{2N} \omega_i^C \left(\mathfrak{X}_k^{(i)-} - \hat{\theta}_k^{-}\right)\left(\mathfrak{X}_k^{(i)-} - \hat{z}_k^{-}\right)^{\mathrm{T}} \tag{77}$$

$$\boldsymbol{K}_k = \left(P_{\hat{\theta}\hat{z}}^{-}\right)_k \left(P_{\hat{z}\hat{z}}^{-}\right)_k^{-1} \tag{78}$$

Finally, the Kalman gain $\boldsymbol{K}_k$ is used to update the state estimation and error covariance:

$$\hat{\theta}_k = \hat{\theta}_k^{-} + \boldsymbol{K}_k \left(z_k - \hat{z}_k^{-}\right) \tag{79}$$

$$P_k = P_k^{-} - \boldsymbol{K}_k \left(P_{\hat{z}\hat{z}}^{-}\right)_k \boldsymbol{K}_k^{\mathrm{T}} \tag{80}$$

9. SIMULATIONS AND NUMERICAL RESULTS

The BT voice transmission system has been created on Matlab/Simulink environments [19]; the demo version has been modified for the purpose of our optimization. We set the transmitting signal power to 10 mW, which is sufficient for an indoor radius of 20 meters. The high level Simulink model is shown in Figure 10 for the BT system [19].

Figure 11 shows the linear KF system model that has been implemented as a cascaded stage at the slave receiver side, and, as it is illustrated, it has two embedded functions: one for imaginary, and the other for real estimation. Furthermore, Figure 12 and Figure 13 will illustrate the modulation scheme for the EKF and UKF respectively, while all of the three Kalman filters have been identified in Figure 6 as a Kalman filter subsystem block.

The simulation results have been analyzed in two different aspects: Bit Error Rate (BER) and Frame Error Rate (FER). The significance of the BER shows the performance of these types of filters in BT system application, and, as a result, it is concluded that the UKF contributes the better result between the linear KF and the EKF. Moreover, the linear KF gives the worst result, which is proof of our expectation about the nonlinear propagation of the transmitting signals, because it is only suitable for linear purposes. According to the theoretical AWGN, shown in Figure 14, the UKF gave the best result at a high noise measurement, and at a low noise measurement, versus the AWGN theoretical result.

Second, in an analysis approach based on the FER, or, in other words, on a package transmitting estimation, we divide every 366 bits into a frame, or a package. The package transmitting concept was more useful on broadcasting data transmission, and, from Figure 15, where the UKF wins the computation of the three types of Kalman filters, according to expectation; the UKF will produce the best results in the matter of the FER. Nonetheless, the FER does not show a theoretical result, but the UKF in Figure 14 illustrates better than do the linear KF and the EKF.

In summary, the experiment simulation is designed and run in different SNR values to come up with the previous results shown in Figure 13, and Figure 14, and each result was computed out of 100 bits of error. Even so, the results will be summarized in Table 1, denoting the meaningful outcomes of this experiment.

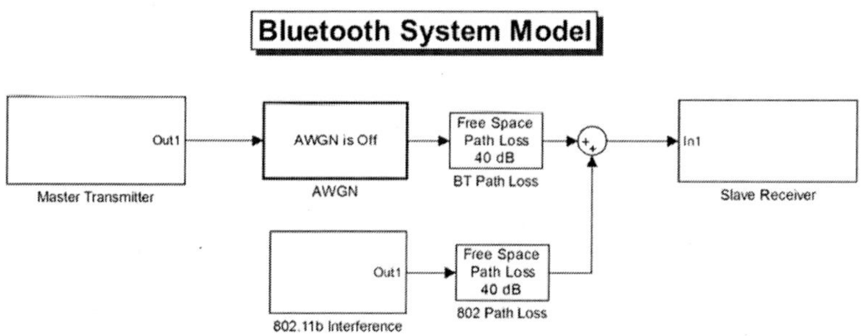

Figure 10. Top level Simulink Bluetooth system model.

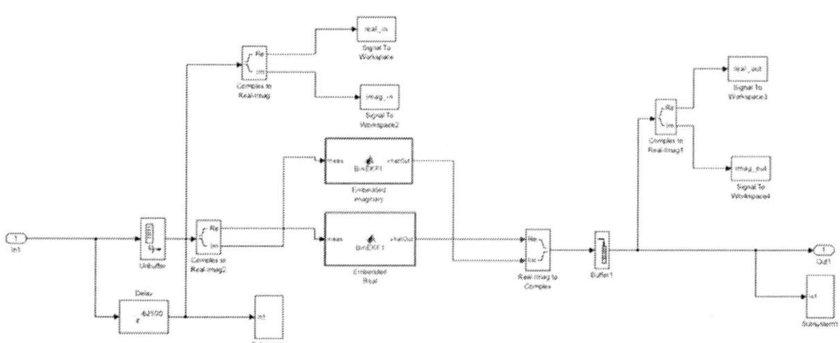

Figure 11. Linear Kalman Filter Scheme implemented at the robust stage at the receiver of the Bluetooth.

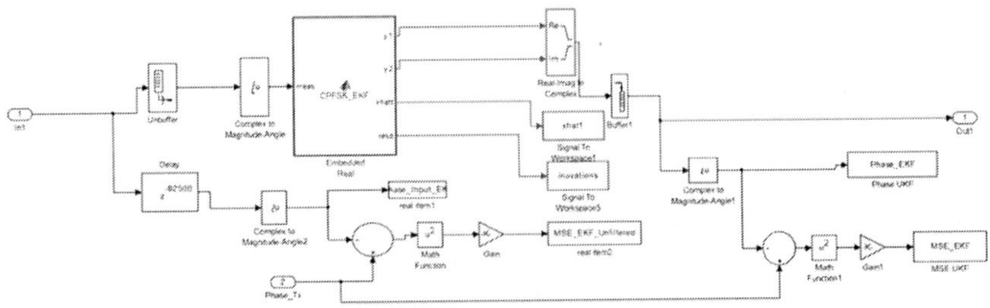

Figure 12. Extended Kalman Filter Scheme implemented at the robust stage at the receiver of the Bluetooth.

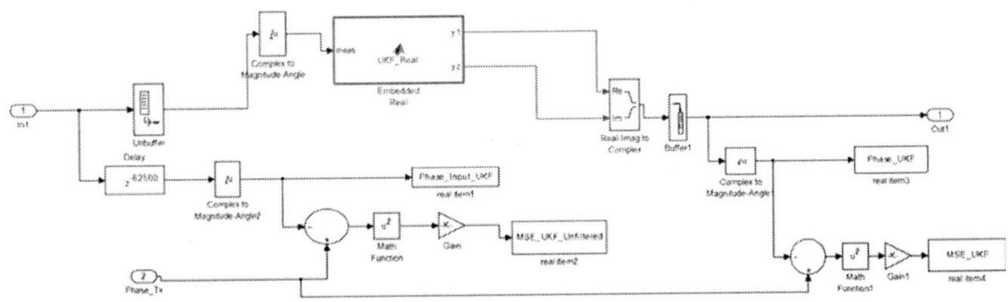

Figure 13. Unscented Kalman Filter Scheme implemented at the robust stage at the receiver of the Bluetooth.

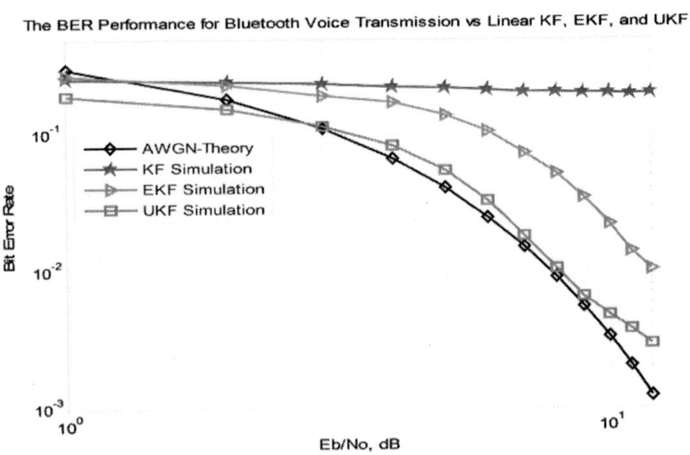

Figure 14. Bit Error Rate (BER).

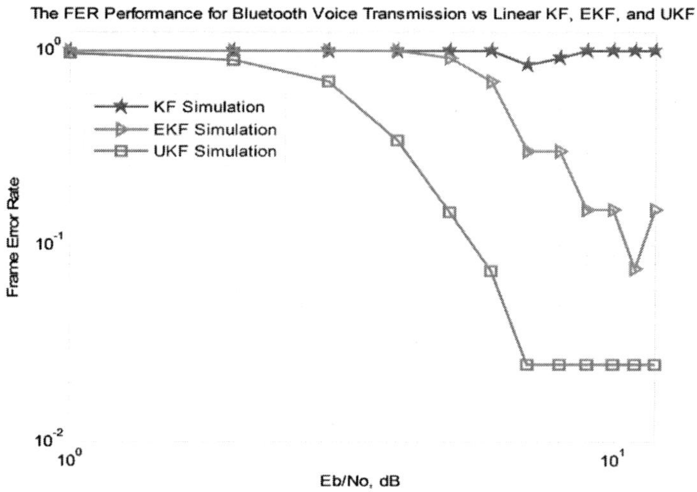

Figure 15. Frame Error Rate (FER).

Table 1. Summarizing the Kalman filters performance verses signal to noise ratio (SNR).

SNR	Linear KF		EKF		UKF	
	BER	FER	BER	FER	BER	FER
1	0.25	1	0.27	1	0.19	0.97
3	0.23	1	0.19	1	0.11	0.70
6	0.21	1	0.10	0.69	0.032	0.075
9	0.20	1	0.034	0.15	0.0065	0.025
12	0.19	1	0.010	0.15	0.0029	0.025

10. CONCLUSION AND FUTURE WORK

In this paper, a novel achievement has been delivered through the design of the UKF, which shows a high qual- ity for the BT system. The challenge was not only filtering a signal, but also utilizing the optimal contributions from among the related works that have been done previously. Therefore, the non-coherent GFSK has been validated using the Bluetooth voice transmission model, which had already been created in Matlab/Simulink software. As a result, an improvement has been executed on the physical layer, which is where we dealt with non-coherent GFSK modulation and demodulation; definitely, our robust scheme has been designed in order to use a faster sampling rate for the execution time of the filter.

There is always room for future works in this area, and some of them are going to be our next tasks, to continue our work. One future task will be applying the same experiment from this paper to an interactive Kalman filter, where the signals have to be modulated into two different techniques simultaneously for tunable purposes. A second task will be to develop the UKF for a multi-path fading channel, which is extensively valid for the 802.11 indoor channel models. The third task will be to consider the multi-user OFDM system, which is useful for master transmitting to Piconet and scatternet; in other words, using a multi-slave receivers connection. Finally, the experiment for the in-vehicle channel mode should be created, which would be a benefit for a tremendous number of the in-vehicle multimedia applications.

ACKNOWLEDGEMENTS

I really would like to thank to Saudi Red Crescent Authority for its help in providing me the time, and the financial support, to establish my Ph.D. research.

I would also like to express my appreciation to the Saudi Arabian Cultural Mission for its infinite support, and supervising us in the United States, in order to achieve our dream.

Additionally, this research paper could not have been done without Debbie Barry, for her editing and linguistics help.

CITE THIS PAPER

Ali S.Alghamdi,Mahdi N.Ali,Mohamed A.Zohdy,11,11, (2015) Robust Non-Coherent Demodulation Scheme for Bluetooth Voice Transmission Using Linear, Extended, and Unscented Kalman Filtering. *Journal of Signal and Information Processing*,**06**,9-27.

REFERENCES

1. B. S. I. Group (2014) Bluetooth SIG. Bluetooth SIG. https://www.bluetooth.org/en-us
2. Leon, I. and Couch, W. (2007) Digital and Analog Communication Systems. 7th Edition, Pearson Prentice Hall, Upper Saddle River.
3. Brown, R.G. and Hwang, P.Y. (2012) Introduction to Random Signals and Applied Kalman Filtering. 4th Edition, John Wiley & Sons, Inc., Hoboken.

4. Nsour, A., Abdallah, A.-S. and Zohdy, M. (2013) Phase Tracking in Bluetooth Receivers Using Extended Kalman Filtering. 3rd International Conference on Wireless Communications, Vehicular Technology, Information Theory and Aerospace & Electronic Systems (VITAE), Atlantic City, 24-27 June 2013, 1-5.

5. Ali, M. and Zohdy, M. (2012) Interactive Kalman Filtering for Differential and Gaussian Frequency Shift Keying Modulation with Application in Bluetooth. Journal of Signal and Information Processing, 3, 63-76.

6. Liu, H., Venkatesan, V., Nilsen, C., Kyker, R. and Magaña, M. (2003) Performance of Frequency Hopped Noncoherent GFSK in Correlated Rayleigh Fading Channels. ICC'03. IEEE International Conference, 11-15 May 2003, 2779-2783.

7. Scholand, T. and Jung, P. (2003) Novel Receiver Structure for Bluetooth Based on Modified Zero-Crossing Demodulation. Global Telecommunications Conference, GLOBECOM'03. IEEE, 1-5 December 2003, 729-733.

8. Lampe, L., Schober, R. and Jain, M. (2005) Noncoherent Sequence Detection Receiver for Bluetooth Systems. Selected Areas in Communications, IEEE Journal, 23, 1718-1727.

9. Roh, J.S. (2006) Performance Analysis and Evaluation of Bluetooth Networks in Wireless Channel Environment. Systems and Networks Communications, ICSNC'06, October 2006, Tahiti, 61 p.

10. Chang, D.-C. and Shiu, T.-H. (2006) Digital GFSK Carrier Synchronization. Circuits and Systems, APCCAS 2006, IEEE Asia Pacific, Singapore, 4-7 December 2006, 1523-1526.

11. Yu, B., Yang, L. and Chong, C.-C. (2011) Optimized Differential GFSK Demodulator. IEEE Transactions on Communications, 59, 1497-1501.

12. Luque, J.R., Morón, M.J. and Casilari, E. (2012) Analytical and Empirical Evaluation of the Impact of Gaussian Noise on the Modulations Employed by Bluetooth Enhanced Data Rates. EURASIP Journal on Wireless Communications and Networking, 2012, 94.

13. Mathew, A., Chandrababu, N., Elleithy, K. and Rizvi, S. (2010) Interference of 802.11B WLAN and Bluetooth: Analysis and Performance Evaluation. International Journal of Computer Networks & Communications (IJCNC), 2, 140-150.

14. John, P.G. and Masoud, S. (2008) Digital Communications. McGraw-Hill, New York.

15. Nsour, A., Abdallah, A.-S. and Zohdy, M. (2013) An Investigation into Using Kalman Filtering for Phase Estimation in Bluetooth Receivers for Gaussian and Non-Gaussian Noise. Proceedings of the 2013 IEEE International Conference on Electro/Information Technology (EIT), Rapid City, 9-11 May 2013, 1-5.

16. Nsour, A., Abdallah, A.-S. and Zohdy, M. (2013) GFSK Phase Estimation Using Extended Kalman Filtering for Non-Gaussian Noise. Proceedings of the Wireless Telecommunications Symposium (WTS), Phoenix, 17-19 April 2013, 1-6.

17. Nsour, A., Abdallah, A.-S. and Zohdy, M. (2014) Non-Coherent Detection of GFSK Using Extended Kalman Filtering for Non-Gaussian Noise. Proceedings of the Wireless Telecommunications Sym-posium (WTS), Washington DC, 9-11 April 2014, 1-6.

18. Mathworks, Inc. (2014) Bluetooth® Voice Transmission.http://www.mathworks.com/help/comm/examples/bluetooth-voice-transmission.html

19. Wan, E. and Van der Merwe, R. (2000) The Unscented Kalman Filter for Nonlinear Estimation. Proceedings of the IEEE 2000 Adaptive Systems for Signal Processing, Communications, and Control Symposium, Lake Louise, 1-4 October 2000, 153-158.

20. Brown, R.G. and Hwang, P.Y.C. (2012) Introduction to Random Signals and Applied Kalman Filtering. In: Brown, R.G. and Hwang, P.Y.C., Eds., Introduction to Random Signals and Applied Kalman Filtering with Matlab Exercises, 4th Edition, John Wiley & Sons, Inc., Hoboken, 141-148.

21. Yang, W.Y., Cho, Y.S., Choo, C.Y., Shim, J., Jeon, W.G., Kim, J., Yu, S., Park, K.W. and Wee, J. (2012) MATLAB/Simulink for Digital Signal Processing. Hongrung Publishing Company, Seoul.

22. Julier, S. and Uhlmann, J. (2004) Unscented Filtering and Nonlinear Estimation. Proceedings of the IEEE, 92, 401- 422.

23. Cho, Y.S., Kim, J., Yang, W.Y. and Kang, C.G. (2010) MIMO-OFDM Wireless Communications with MATLAB. John Wiley& Sons (Asia), Singapore.

Index

I

K

M

N

P

R

S

U

Z